The Complete Book of Willows

S caprea, the palm or goat willow at woolly seeding stage (*M. Nimmo*)

The Complete Book of
Willows

S. C. Warren-Wren

South Brunswick and New York:
A. S. Barnes and Company

First American edition published 1973 by A. S. Barnes and Co., Inc., Cranbury, New Jersey 08512

Library of Congress Catalogue Card Number: 72–5878

ISBN 0–498–01262–X

Printed in Great Britain

Contents

Contents

Illustrations

LINE DRAWINGS

by John Green

Preface

'Only God can make a tree'. Most of us have long accepted and respected this statement and over the centuries we have striven hard to widen our knowledge and understanding of the vast arboreal scene. Such study has revealed intensive problems, certain genera being more troublesome than others. The willow being heterogeneous and highly disposed to cross-breeding has tended to be among the worst offenders, being ready at all times to confuse, to confound and to create disagreements in plenty among the great and learned men. In fact, the genus is said to be the most difficult and inextricable in the vegetable kingdom.

Be that as it may, an offer of a complete solution to all the problems is not the purpose of this book. Rather, it is to reveal the immense possibilities of a fuller utilisation of the genus in all its many forms. Despite the fact that the genus is among the earliest on record many of its attractive qualities are as yet unknown to the general public and, indeed, to many horticulturists. It can even be said that willow is the Cinderella of the arboreal and horticultural scene, that it has waited over an æon of time for a fairy godmother to grant her wish to go to the ball. As it has been my privilege to work very closely with willows over many years and to study in some detail their rather unusual and, in many instances, distinctly attractive characteristics I feel able to record something of their true significance and potentialities. The genus has the capacity to combine *utili dolce* to a degree where it becomes capable of reaching, or even surpassing, many other more esteemed plants and, let us say, going to the 'ball' on equal terms with the rest.

The 'superior' numbers appearing throughout the text, usually at the end of the sentence, relate to their corresponding numbers in the list of References appearing at the end of the book.

Measurements of plants are given in the metric system and where the English figures appear in brackets, these are 'rounded off' figures and are given solely as a guide to readers and not as exact equivalents in metric measurements.

S. C. WARREN-WREN

1 : *Introduction*

EARLY DAYS

THE WILLOW IS as old as any of the earliest recorded plants. It is definitely of pre-Ice Age and, with the exception of a few insignificant equatorial regions, it has practically a world-wide distribution.

Geological history records three major glacial periods and during each of these a relatively small part of the earth's surface was covered with ice, which gradually retreated northwards and finally disappeared. The exposed areas at the end of each period consisted largely of great lakes and meres with raw water-logged ground in the low-lying parts allowing any surviving trees and woody plants to settle and spread.

The origin of the willow is most conveniently traced by turning to the rocks and fossils, reconstructing and piecing together the facts which they contribute and drawing conclusions therefrom. Evidence produced by pollen grains in fossils has shown the dominant vegetation to consist of dwarf birches, sedges and arctic willows. The first plants in the fossils discovered during the second and third glacial periods consisted mainly of plants of the group known botanically as Angiosperms, to which willows belong. That was about 200 million years ago.

The fossils of the Cretaceous system reveal for the most part pollen grains and the remains of leaves. Fossils of other and later periods have included willow roots and stems but, unfortunately, few offer any conclusive evidence of the type or variety of the willow. It is most probable that the remains were the produce of the dwarf or arctic types but it should not be assumed that the willows present in those far-off times were either few in number, restricted in form or that they were not descendants of even more ancient ancestors. In drawing conclusions from a study of fossils and rocks we have not to overrate the accuracy of the results, as assumption and hypothesis are bound to play an important part. In fact, it is generally believed that we have not, in a fossilised condition, a fraction of the plants that have existed, and that not a fraction of these we have are recognisable specifically. Therefore, the final results can scarcely be other than guesses at the truth—a truth, so far as willow is concerned, being based on a limited but unbiased knowledge and the only information available.

Peat taken from an area between the Alps and the Jura mountains at a height of 488m (1,600ft) above sea level, revealed evidence of several species of willow as being present during the period immediately following the close of the Pleistocene Ice Age. Even in the unglaciated areas one willow, *S.brachycarpa* Nuttall occurred in British Columbia, Washington, Oregon and then 2,414km (1,500 miles) distant in Gaspé County, Quebec. There is definite evidence that this, and other flora of these unglaciated areas of borial America, are survivals of the pre-glacial ages. Above the Cromerian series of the East Anglian pliocene strata is a band of clay known as the arctic plant-bed, so called because of the occur rence there of *S.polaris* Wahl., the arctic willow. Today this species appears commonly in Spitsbergen and district.[14]

An analysis of peat bogs situated in various regions, dating from the post-glacial periods, reveals much evidence of the presence of willow deposited either before or after the ice had finally melted. Charcoal from the Neolithic and Iron Ages both early and late has also produced evidence of the presence of willow during those respective periods. The first willow of prehistoric times to be specifically named appears to have been *S.repens* L. or the creeping willow. Its fossilised remains originated in sand dunes and marsh brushwood in south-eastern England. There is evidence too of the existence of willows very shortly after the close of the Pleistocene period. These are believed to include *S.herbacea* L. and *S.reticulata* L. both species being among the earliest members of the Angiosperms to recolonise after the retreat of the ice.

From time immemorial the willow in its wild state has been, and still is, among the first vegetation to appear on ground laid bare after the extraction of gravel, war-time devastation, earthquakes and burned-over forest areas. One particular species, *S.bebbiana Sarg.*, native to the eastern United States of America, also Canada, commonly provides this type of public service on the burned-over forest areas of these regions. There is little doubt that insistence on being at the forefront of vegetative growth on barren sites is matched only by its tenacity to hang on to life under the most severe conditions, for willow is certainly one of the most difficult plants to destroy by any means other than fire.[2]

WORLD-WIDE DISTRIBUTION

It would appear that the ability of the willow to adapt itself to diversified climatic conditions has enabled it to overcome the apparent neglect of man to recognise its great potentialities. Throughout the years it has increased naturally and spread over vast geographical areas. So far as can be ascertained the only parts of the world where willow has not been found are the Malay Peninsula, Polynesia and a few tropical islands in the equatorial belt. The absence of willow from these regions could be due to one, or both, of two causes; either the plant has not been introduced into these areas, naturally or by the efforts of man, or, that the persistently high temperatures and lengthy periods of drought have prevented the growth and development of any plants which may have reached those parts. After all, the prevailing climate thereabouts is quite inconsistent with the requirements of the genus.

Willow is to be found in every continent but it is most prolific in the arctic, alpine and north temperate zones of Europe and America. Many species flourish over wide areas of central, north and eastern Russia and southward to many parts of Asia including India, Java, Sumatra, the Philippines, Egypt, and several parts of northern Africa. It is present in the vegetation of the following islands: Iceland, Greenland, the Canaries, Madeira, Madagascar, Ceylon and Newfoundland. While there is no evidence to show that willows are indigenous to Australia, small numbers of the *fragilis* and *alba* types are to be found along the banks of the larger water courses of New South Wales and Victoria.

As with northern Europe, so with the arctic and temperate zones of north America and Canada the willow is prolific in its numbers and varieties. It flourishes southward throughout the United States and in lesser degree beyond the Equator into South America. On the whole the species and varieties in these parts differ very slightly from those of Europe and Asia. Despite the fact that the Americas comprise the so called New World, most of its willow population can be traced back to pre-Ice Age. For a very long time the osier willow varieties have been cultivated there for the manufacture of baskets and other products; a good number of ornamental willows are to be found in the private and public gardens widespread throughout the Continent, and exchange of varieties between the continents by private individuals has been made in a number of instances. In fact, the distribution of willow is not dependent alone upon the movement of seeds through the air and water, but a not inconsiderable measure of exchange has been made by the passing of cuttings and plants between growers in the various regions.

In the nineteenth century, 1817 to be precise, a Dr. Barratt of Middletown, Connecticut, USA, set out to describe all the willows of North America, indigenous and exotic, his conclusion being that they amounted to about 100 different types. Cuttings from most of these were sent over the Atlantic to the then Duke of Bedford, who planted them in his Salictum at Woburn in Bedfordshire. Many of these were re-distributed to the Duke's many friends in Britain and in Europe.[8]

Another instance of long distance distribution occurred some years ago when a well-known willow merchant named Watts despatched a number of willow sets of the coerulean variety to Queensland, Australia. While the majority of these are still believed to be alive their slow rate of growth is such as to force me to the conclusion that the Australian climate is unsuitable for commercial willow, for which a fast rate of growth is essential. Another instance of export of coerulean willow sets abroad was undertaken about eight years ago when I despatched a small consignment to Algeria. At the time of writing I have no up to date information concerning the health or progress of these trees.

Where the willows enumerated under various sub-headings in Chapters 3 and 4 are known to be growing in north and south America and Canada an indication of this is given in the entry. It would seem, however, that there are a small number of willows recorded as being present in America, but for which I can find no record on the eastern side of the Atlantic. These are named below:

S.*altrissima* is a tree attaining a height of 120ft and found in Arkansas, Louisiana and Texas.

S.*amygdaloides* Anderss. Western Quebec to British Columbia, along the Great Lakes provinces, south to central New York, westward to the Rocky Mountains and southward along their slopes to north west Texas, Oklahoma, New Jersey, Kentucky and Nebraska. Not common in the east.

S.*discolor* Muhl. Native of eastern USA and Canada, Nova Scotia to Manitoba, south to Delaware and Missouri.

S.*eriocephala* Michs. (related to *discolor* Muhl.). Widespread over the American continent. Has very hairy twigs, leaves often permanently rusty, hairy beneath and coarsely toothed.

S.*falcata* Jorr. (a form of *nigra* Marsh). Common from Massachusetts southward.

S.*harbisonii* Schn. Found on the edges of streams and swamps from Virginia to Florida.

S.*missouriensis* Bebb. Found in the sandy river bottoms of the tributaries of the Missouri and Mississippi rivers from Kentucky to Illinois and Nebraska.

S.*pedicellaris* Anderson. Native of eastern N America. (? related to S.*myrtilloides* L.)

S.*scouleriana* Barr. Found from Alaska to California, eastward to Idaho and Montana. It is the most abundant willow in the Pacific states.

Page 17 (right) A *fifty year old pollard with little or no body but a well maintained and virile crown, that is the hybrid* S. Albescens (alba × fragilis)

(left) *the 'Crack' willow* (S. Fragilis L) *in early spring*

Page 18 (right) *Leaf sprays of the 'bay leaved' willow, lustrous dark polished green, and aromatic when crushed;* (below) *the bole and bark of the attractive 'bay leaved' willow* (S. pentandra)

2: *Specific Willow Features*

TYPES, FLOWERS AND SEEDS

WILLOWS CONSIST OF three typical forms—upright trees and bushes, pendulous or weeping types and dwarf or trailing plants. They are arborescent and herbaceous and, in general, are termed woody perennials, that is, plants having stems or portions above ground enduring for longer than one year. When the large timber-bearing trees are in cultivation they are regarded as horticultural plants or crops; when growing in their wild state they rank as forest trees. As in many other aspects botanists have differed in their opinions regarding the influence of sex upon the character and appearance of willow. There is little doubt that in certain species differences do occur as between the male and female plants, particularly in regard to the foliage. In other species any differences there may be are scarcely distinguishable. The white willow, *S.alba L.*, is an excellent example of where the sexes show no difference in foliage while in a few other respects small variations are discernible.

The leaf and flower buds are truly wonderful works of nature and when bursting time arrives it is difficult to believe that their development process has been going on for many months. This period of their life is not without its dangers; drought combined with insect pests appear in plenty with the arrival of the summer months, and frosts and snow and hungry birds are never far away with the onset of winter. All these are intent upon taking their toll of the attractive developing life. Mother Nature having created only a delicate bud-structure, consisting of undeveloped shoots, makes amends by providing several different devices to aid the plant's survival throughout the changing seasons. The protective coverings vary slightly according to the species or variety of the willow. In general, they assume the character of a tough scale with hairy or furry coverings impregnated with resins and gums capable of keeping out the rain and preventing the internal moisture from escaping.

Willow flowers are borne mostly on erect catkins, only rarely do they adopt a spreading or even a drooping attitude. (See *S.basfordiana* Scaling, for example.) They are of the simple form, lacking petals. While a certain degree of variation occurs as between the many differing kinds of willow, in general, the male, or staminate flower, consists of two stamens and one, two, or more nectaries borne in the axil of a hairy scale. The female or pistillate flower is comprised of bract, ovary, and style, ending above with a forked stigma to receive the pollen. The stigma is separated at the top, the two halves curling themselves in opposite directions exposing a mass of silky seeds. The nectary in each bract exudes a sweet juice—hence the attractiveness of the flower to insects. A vast number of female flowers is produced, each spike bearing a great quantity of seeds. The size of the seeds is naturally in an inverse proportion to their number. When the bud-scale falls in early spring it exposes a catkin which may be practically stalkless, short-stalked or even long-stalked. It is densely packed with flowers which are almost invariably unisexual.

Incidentally, flowers of both sexes secrete honey and as most willows bloom very early in the year, certainly well in advance of most other honey-producing flowers, they are not only highly prized by beekeepers but prove very attractive to insects, honey bees, their cousins the wild bees, and also to wasps. In addition to collecting the honey they help to spread the pollen. After pollination has been effected, the male catkins wither and drop, while those of the female become more noticeable, the ovary tending to increase in size. Both bees and wasps frequently take up residence in the trunks of the older willows. I have seen as many as three colonies of bees and one of wasps all sharing, presumably amicably, one arboreal residence.

In its wild state willow is distributed naturally both by the dissemination of its seed and by vegetative growth. It is essential that the seeds of most kinds of willow find suitable conditions in which to germinate almost immediately upon release, certainly within one week or less, or they lose their vitality. In fact, it is because such conditions are not always immediately available that only a small fraction of the vast seed production actually propagate. Since the seeds are dispersed mainly by the wind they must be light and, as many will be lost, they must necessarily be numerous. Each seed bears a tuft of hairs which open out like rays as they are released. As they become dry they are borne on the wind for great distances. Seeds of the arctic species are blown on to and across the ice packs thereby spreading the seeds widely between the islands.

During the summer months the ground around female trees of several species is often white with a wool-like substance which is none other than the seed-carrying medium having fallen from the trees. The woolly hair of some willow seeds is sometimes freed from the capsules without seed being attached, resulting in the distribution being all to no purpose. This could be due to there being no male plant close enough to fertilise the ovule and thus enable it to develop. Seeds of certain willow species have been known to float in the air for one and a half days while others, chiefly mountain species, which have greater buoyancy, stay in the air for as long as 2½–3 days. The speed of fall of the plumed willow seeds is also of interest. Three examples are:

S.repens L. falls 3·66m (12ft) in 22·6 seconds
S.aurita L. falls 3·66m (12ft) in 19·4 seconds
S.pentandra L. falls 3·66m (12ft) in 15·2 seconds.

These figures relate to the rate of fall in still air. If a breeze is blowing, the rate of fall will be correspondingly lower, while the distance travelled before the completion of the fall will be greater. Many instances are on record of birds carrying the plumed willow fruits and seeds considerable distances for use in the construction of their nests. On several occasions I have found them in hedgerows, the seeds having fallen from partly constructed nests of both blackbirds and thrushes. In such instances the nearest willow has been not less than three miles distant. An interesting medium of seed distribution is provided by the rough coats of sheep, cattle, goats and horses, to which the hairy fruits and seeds become attached as the animals pass through the ground where willows are growing.

SALLOW PLANTATION

The sallows are a distinct group of willows consisting of large bushy shrubs and small trees. They are readily identifiable by their leaves which are almost invariably obovate or rounded, shaggy and thick and early appearing. Their catkins are silken with prominent yellow and distinct stamens, two to a flower. Approximately forty different sallows are to be found widespread over many international locations including east and west Europe, Asia, N Africa and certain areas of the USA including the marshes of Pennsylvania.

The majority of sallows have a strangely modest yet not unattractive appearance. As so few of them have been brought into cultivation little is known of their cultural needs or possibilities, though I doubt if their needs are finical to any large degree. In their wild habitats they keep remarkably free from pests; they thrive best in poor quality soil and require very little water. Any attention one felt like affording to sallow should be kept to a minimum—they prefer it that way. They resemble a group of natives brought out of their wild environment, and who, if treated with an excess of civilisation, could die under the strain! Sallows, if over cultivated, could react similarly. Anyone possessing a rough, not necessarily damp, piece of land, otherwise unproductive, may like the idea of planting it with sallow. They are not unbeautiful plants and the mere fact of congregating them into one place may produce interesting results. Cuttings of sallow types are not difficult or expensive to obtain. In fact many varieties can be gathered free of cost for the cutting, from hedgerows, gravel pits and small woodlands, always, of course, with the property-owner's permission. Details follow of forty-one different sallows:

S.acuminata Sm., or the long leaved sallow, is a tree of considerable size for a sallow. It flowers in early April; catkins resemble somewhat those of *caprea* but are longer and not half so thick.

The green mountain sallow *S.andersoniana* Sm., has leaves of rich bright green during the first summer but in following years they turn to sooty brown which, strange as it may seem, appear quite beautiful.

Anson's sallow, *S.ansoniana* Forbes, is a spreading bushy shrub with dark mahogany coloured branches, glabrous and shining after the first year. Its leaves are bluntly and deeply serrated. Not in the upper class maybe, but not to be despised either.

The water sallow, aptly named *S.aquatica* Sm., is naturally found in particularly wet places. It is generally bushy and rarely forms a tree; branches are numerous and upright, brittle and not adapted to any economical purpose.

The dark-purple-branched sallow *S.atropurpurea* Forbes, has branches covered with minute hairs, marked with small yellow spots and very brittle.

S.atrovirens Forbes, is the dark green sallow and a native of Switzerland. It is an upright shrub or small tree up to 4m (12ft) and flowers in May.

The round-eared or trailing sallow *S.aurita* L., is a shrub up to 9ft in height with numerous spreading branches. Catkins appear before the leaves, male are ovoid and the female cylindrical. Numerous in damp woods, heaths, rocks by

S.aurita, the wrinkled sallow; leaves and male catkins

streams, and on moors; favours light acid or slightly basic soils, ascending to 867m (2,600ft).

S.australis Forbes, is the southern sallow and flowers in April or May and is a low, upright, bushy shrub with reddish-brown, downy branches. It is not outstanding in any way, but just one of a number which when together look quite good.

S.caprea L., is a shrub, or may be found as a tree up to 13m (39ft) and is locally known as 'the palm', 'the great sallow', 'the goat willow' or 'great round-leaved sallow' and, of course, the 'pussy' willow. The female bears large silvery catkins in very great profusion. On a late evening the copse or hedgerow may be lit by the yellow of the male catkins. The leaves are softly silky to the touch, are deep green above and softly grey beneath, broadly oval or almost rounded. The wood is white, tough and smooth and is used for making charcoal, gun-powder, tannin, salicin and a number of other products are also taken from it. It is a prolific grower, making itself at home in any soil in woodland and hedgerow, rather than by water. It makes an excellent underwood for coppices. Bees are attracted to it in large numbers by its lavish provision of pollen and nectar. It may be useful to place some hives in or near the sallow plantation.

As if not to be outdone, the sallows have a pendulous variety to add to their list. It is *S.caprea pendula* Th. Lang, or the weeping sallow. It is not outstanding but I am sure it would give a pleasing effect if planted among its fellows in the sallow field.

S.carpinifolia Schl., is a small bushy tree with round villous branches of a sooty brown colour, with hairy buds and catkins one inch long. Commonly called the Hornbeam-leaved sallow it flowers in March and April and is just another sallow. It needs to be sited close to sallows possessing more attractive qualities, to give it some significance.

The grey or common sallow *S.cinerea* L., is a shrub or more

S.caprea, the Goat willow; leaves and male and female catkins rarely a small tree up to 9m (30ft). It is found mainly in the fens of eastern England and often dominant in carrs, damp woods and other wet habitats.

The hybrid *S.coriacea* Schl., is the coriaceous-leaved sallow which flowers in March. It is a bushy shrub of 2–3m (7–8ft) with pale green branches remotely marked with yellow spots. The substance of the leaves is of a thick leathery texture.

S.crassifolia Forbes, is the thick-leaved sallow, a bushy shrub about 3m (10ft) high with dark green, downy, brittle branches, soft to the touch when young. The substance of the leaves is thick and coriaceous. Catkins appear before the leaves, ultimately 5cm (2in) long.

The damson-leaved sallow, *S.damascena* Forbes is often confused with several other varieties but the leaves and twigs of this plant closely resemble the damson plum; the stem and branches are erect, dark brown-mahogany colour, with small yellow spots, round and brittle.

In *S.dura* Schl., the rapid growing sallow, we have a departure from the normal in sallows. On the whole they are not fast growers, at least, not in their wild state; what will happen when they are moved into cultivation, or partial cultivation, is a matter of conjecture.

The ferruginous-leaved sallow *S.ferruginea* Anderss., is a bushy shrub or low tree up to 4·3m (14ft) in height, with shortish, green, fuscous branches, round, down of a rusty hue towards autumn but pale yellow earlier.

S.firma Forbes, is the firm-leaved sallow, a straggling bushy shrub. Specimens in a garden flower in March and April and again in August. It would be interesting to see how many other sallows would give us two flowerings a year if and when brought into cultivation.

The glaucous mountain sallow, *S.forsteriana* Sm., has an erect stem, minutely downy branches, and flowers in May. It was found on the Breadalbane mountains of Scotland.

The hybrid *S.geminata* Forbes, or the twin-catkin sallow,

is a native of Britain. It has large catkins, often more than one bursting from the same bud. It is another of the rapidly-growing types of sallow and flowers in March.

The Grisons Sallow *S.grisonensis* Forbes, is a native of the Grisons, in Switzerland, and was introduced to Britain in 1824. It flowers in March and April.

The grey-leaved sallow *S.grisophylla* Forbes, is a shrub up to 1·8m (6ft) in height and is to be found wild in the marshes of Pennsylvania and other parts of the United States of America. Its branches are brown and downy when young; leaves are 3·8cm (1½in) long; catkins 2·5cm (1in) long, obtuse, and appear before the leaves. Introduced to Britain in 1824.[12]

The Swiss sallow *S.helvetica* Forbes, is a bushy tree, very distinct in form of leaf and mode of growth. Up to 4m (nearly 14ft) in height; with greenish-brown villous branches copiously marked with yellow dots. Leaves up to 5cm (2in) in length and about 3·4cm (1½in) broad, silver-grey in colour. Catkins are about 2·5cm (1in) long, the males are yellow and in great profusion.

S.hirta Sm., is the hairy-branched sallow, a small tree and remarkable for its thick round hoary branches, clothed very densely with prominent, close, horizontal, soft, cottony hairs.

S.incanescens Schl., the white-leaved sallow is a bushy shrub or tree with round branches, muddy-green colour, marked with a few yellow spots as though besmeared with clay. Catkins appear before the leaves in March, with a second appearance in August.

The lake sallow *S.lacustris* Forbes, is a straggling shrub with round, dark, pendulous branches; leaves serrated and dull green; catkins are 2·5–3·8cm (1–1½in) long. It flowers in March.

The hybrid, *S.latifolia* Anderss., is a straggling sallow with leaves of large broad, elliptical form, green and shining above, glaucous and downy beneath, finely serrated at the apex. It flowers in March and catkins are nearly 2·5cm (1in) long at maturity.

Another rather fast growing sallow is *S.macrostipulacea* Forbes, (x stipularis); it has dark green, round downy branches, marked with small yellow or reddish spots, the lower branches being pendulous. The young leaves are purplish, soft to the touch and pubescent. Adult leaves are rather coriaceous, copiously marked with dark blotches. Flowers April and May.

S.mutabilis Forbes, or the changeable sallow, flowers in March and April; branches are densely downy, copiously beset with somewhat elliptical leaves which are a dark green colour above, pale and hairy beneath, with prominent veins; their substance is rather of a thin crackling texture; young leaves very hairy.

S.oleifolia Macreight, (a ssp of cinerea), the olive-leaved sallow. An English native, flowering in March and again in August. It is distinguished from *cinerea* by the coriaceous texture of its leaves and it very much resembles *Quercus Ilex*.

S.pannosa Forbes, or the cloth-leaved sallow is a tree up to 3·6–4·2m (12–14ft) in height, oblique spreading branches of darkish, fuscous colour; young twigs greyish-brown and downy. Many of the leaves are alternate and opposite on the same branches. Catkins about 2·5cm (1in) long.

The rock sallow *S.petraea* Anderss., flowers in May. It is a British shrub up to 4·5m (15ft) in height, with crooked ash-coloured branches and brown twigs. Leaves are long, very dark green, shining on upper surface; bluish, rather more hairy and woolly on under side.

S.rivularis Forbes, is the river sallow. (It is often referred to as 'rivalis'.) It is an erect-growing shrub with dark mahogany-coloured branches, nearly perpendicular in mode of growth, copiously marked with yellow dots, young ones green and pubescent. Leaves 2½–3cm (1–1½in) long with short, oblique points, dark green on upper surface; catkins obtuse and short.

S.rotundata Forbes (resembles *carpinifolia* Schl.) Flowers

S.cinerea, var. *oleifolia*; leaves and male catkins

April and May. An upright-growing shrub or low tree up to 4·6m (15ft) or slightly more. Leaves orbicular, bluntly serrated, glabrous and shining above; glaucous, reticulated and minutely hairy beneath, almost glabrous at maturity. Catkins 2·5cm (1in) long.

The silky rock sallow *S.rupestris* Down., flowers in April. Stem is trailing and depressed, dark coloured branches covered with very fine down when young. Leaves about 2·5cm (1in) long, obovate or elliptical. Catkins appear rather before the leaves, 1·3cm (½in) long.

S.schleicheriana Kerner, is a sallow that flowers in April. A bushy upright-growing shrub with yellow, round, pubescent branches variously marked with small black spots. Buds yellow. Catkins, which appear before the leaves, all incline to one side of the branch and are very numerous.

The sordid sallow *S.sordida* Kerner, flowers in April. An upright-growing bushy shrub with yellow, round, pubescent branches, variously marked with small black spots. Buds yellow. Catkins before the leaves, numerous recurved. Anthers yellow. Filaments whitish. Twigs brittle.

S.sphacelata Wahl., is the withered-pointed-leaved sallow and is a bushy tree up to 2·4m (8ft) in height. (It is a variety of *caprea.*) Young branches very soft with dense, hoary, short, velvet-like down. Leaves also soft and downy, greyish, obovate or elliptical; upper side light green with fine down which disappears, underside more downy. Tip of leaf looks as if blasted or withered and assumes a tawny hue. Catkins 3·8cm (1½in) long when mature. Stem erect.

The creeking sallow *S.strepida* Forbes, flowers in March and April. A straggling bush with long pendulous branches of pale greenish colour, pubescent, soft to touch, perfectly round. Buds purplish and hairy. Leaves about 5cm (2in) long, broadest about the middle, obovate-elliptical, the tip oblique.

The elm-leaved sallow *S.ulmifolia* Forbes, is an attractive bushy tree up to 5·5m (18ft) or more. Stem erect. In cultivation, flowers appear during April and again in August.

The vaudois sallow *S.vaudensis* Schl., flowers in March and April. A low spreading bushy shrub, slender, round, downy branches, reddish at first, becoming dark, sooty brown colour after first year. Young leaves are purplish on luxuriant shoots.

SALICTA WITH A PURPOSE

At various periods a number of persons have built up collections of willows with the object of studying more about them, botanically and culturally. Unfortunately, many of these have developed into half-hearted affairs, not lasting sufficiently long in an active state to achieve very much in the way of education or research. At the present time the most effective collections are to be found in the botanic gardens of Britain, Europe and America, but even in these there is one general failing which tends to deplete the usefulness of an otherwise excellent idea. It is the insufficient space allocated to the willows, in other words, overcrowding of the plants and thus preventing them from revealing themselves for effective study. There is so much more information still to be gathered concerning the genus that facilities for one or more salicta, capable of overcoming the present shortcomings, should be established.[15]

A salictum should be a sizeable area of ground set aside for the express purpose of growing a fully representative group of willows. The purpose of the project is to enable study and research into the many peculiarities and problems of the genus.

Provision should be made for as many as possible of the various characters of willow plants to be revealed; the site should consist of a variety of soils and water tables, be wholly aquatic in parts and rocky, gravelly, sandy or arid in other parts, with localities of heavy rich loams interspersed among them. In the course of planting, the willows should be grouped in various ways—on rocks, banks and stony places; open countryside, including upland and valley sites. It may be considered wise to have two salicta—one in the reasonably cold, mountainous regions of, say Wales, Scotland or northern North America, and a second situated in a fertile, level area in part of a temperate zone, capable of being developed on gardenesque lines.

If water is at command it may either be conducted in drains under the surface and tapped for use as required, or it may be present in the form of a canal or river abutting on to one side of the area involved; or it could be in the form of a lake within the area of the salictum with a number of inlet and outlet ditches arranged in geometrical form. Naturally, the willows would be sited in the most appropriate places for the maximum use to be made of such facilities and for their beauty to be revealed to the fullest extent, bearing in mind the presence of the water. It would be well, in such circumstances, to avoid all temptation to interfere in any way with the natural style of growth of the willows, according to their respective types, eg grafting any member of the creeping or trailing types standard high, with the object of changing the effect, should not be done.

Most willows flower in early spring up to mid summer. In the daytime, when the sun shines, the flowers are normally covered with many bees, partaking of the nectar so lavishly provided. In such circumstances it appears reasonable to include some beehives in the paraphernalia of the salictum, enabling it to become one of the most cheerful and inviting of garden scenes. Apart from its scientific value to botany and research the salictum could take its rightful place with the delightful arboreta placed strategically over many regions and, like them, be opened for the public enjoyment and edification.

In the event of a salictum not being close at hand there is nothing to prevent a gardener or a forester from stepping out into the countryside, taking heart and a sharp knife, and gathering specimens of willows as they are encountered, later comparing them with illustrations and descriptions in tree manuals, or, better still, with herbarium specimens of known origin, where this is possible. A wiser course, perhaps, is not to go alone but to take someone along who knows the answers—or most of them—and who is patient and generous enough to pass them on. To aid the accuracy of the exercise, specimens should be taken during at least three different periods of the year: (i) when the plant is in flower (ii) when in early leaf, capsules ripening and stipules in perfection and (iii) when leaves have attained their full size and consistency. Because of the necessity for subsequent visits, the plants from which the first specimens were taken should be marked to ensure that they are easily recognisable on future visits.

EPIPHYTES IN THE WILLOW TOPS

The willow is practically unlimited in regard to the situations in which it may be found; gardens and parks, meadows, woodlands, mountains and valleys and heathlands—all offer something for one or many of the varied members of the genus in which they may find satisfaction with life.

That a plant can show gratitude for what nature has made available to it, is scarcely given a thought. But when something akin to returning favours occurs one must take note of it. In this instance the act takes a rather intimate form and effects the life and development of a mass of flora and fauna through the provision for them of practically ideal habitats. The return favour, so to speak, is the offered use of the crowns of the older lowland riverside willows, which commences when young trees were sawn off at about 8–9ft from the ground. New shoots grew from the cut tops, out of reach of cattle and were permitted to grow for 12–15 years when they were cut for farm fencing purposes. A number of crops of

Page 35 (left) This beautiful willow, S. tristis, is known the world over and rapidly replacing the once favourite weeping willow, S. babylonica L; (below) S. humilis, the Prairie willow, is a native shrub of North America with attractive catkins, stigmas and anthers, in a colour scheme of red and yellow

Page 36 *The strong and very hardy weeping hybrid,* S. sepulcralis, *growing wild in Blowering Valley, near Tumut, New South Wales, is one of the very few willows to be found in Australia*

such poles were taken over the years but, today, little of this economic proposition is exploited. The poles are no longer cut from the tops, the earlier pollarded willows being allowed to expand into wide open crowns with basin-shaped depressions in the middle.

The depressions are capable of collecting and retaining fallen leaves, used birds nests, decayed remains of mosses and lichens and other fertile matter carried on the wind or by other means. In time, a form of woodland leaf-mould is built up in which seeds of many kinds find a resting place in conditions satisfactory for their survival and germination. Plants developed in this manner are known as epiphytes and while certain species are more commonly found than others the variety is quite large. The height of the crown above the ground forms an ecological barrier allowing no plants to pass save those with the means and the opportunity to do so.

Not only the flora and fungi are accommodated in this aerial homestead, but it can accommodate an astonishingly rich assemblage of animals, earthworms, which help to render the dead leaves and dust into a rich mould, woodlice, millipedes, spiders, beetles and the like; shrews come to stay and feed on the insects; hawks and owls prey on the shrews and tits, and so a food chain soon comes into being. It is of interest to note that epiphytes, unlike parasites, ask nothing more of their hosts, the willows, than the use of the 'open crown' as an accommodation area, which, when once occupied can be retained long after the death of the willow, even until the stump eventually rots and falls.

3: *Willows for Gardens and Parks*

ORNAMENTAL TREES AND LARGE SHRUBS

IT IS SCARCELY possible to divide willows into watertight compartments, so readers may find certain varieties occasionally appearing in two or more of the groups into which they are now being allotted to create some semblance of order. Because of the great variety of forms, willow plants are found suitable for practically all types of growing areas. Gardens, large and small; rockeries; parklands—private and public; low-lying watermeadows, and those odd corners of damp land unsuitable for any other purpose are all suitable situations for one or other of the willows.

If we wish to increase the scope of our planting we may utilise some of the lesser known types and with them create a salictum or willow field. It may be that we shall thereby get to know them all very much better and so widen the range of plants for our pleasure. The genus has been neglected for so long that there is ample scope for such planting and study. Incidentally, I was speaking to a gardener recently and told him I was writing this book with the general idea of giving the willows a fillip, and he retorted, 'I don't know what light you can throw on that drab old lot!' Well, let us here and now reveal the brightness of a goodly number of that 'drab old lot', taken at random from our lists. Many, but not all of the ornamental willows have most to offer during the winter months or in very early spring. That is greatly to their credit, for is it not during the winter months that we welcome all the brightness we can find? Moreover, the delightful effects of the colourful buds, stems and flowers, are enhanced still more when set against backgrounds of glistening snow.

There are willows with stems of bright yellow and various shades of red; young shoots of brilliant orange-scarlet; golden-yellow shading into orange scarlet, sometimes in the upper half of the leaf only, with the remainder bright green; narrow slender leaves opposed to others quite broad, close set and silky; small round-headed trees with silvery-white leaves; tall timber trees, pyramidal in shape, with grey-green foliage and yolk-of-egg yellow young shoots. Yes, it is truly a list of attractive things—with more to come; for instance, the sombre-sounding, strangely-beautiful tree with black and heavily-furrowed bark. Set this against another with fine spreading branchlets loaded with long, grey (female) and pink and orange (male) catkins, which appear in the early spring—the comparison is impressive. A particularly attractive Chinese shrub-type crowded with conspicuous catkins in October; long, rosemary-like, narrow hoary leaves borne on bamboo-like wands, with plumose effect; a beautiful Asiatic tree made still more attractive by its twigs and branches being much twisted and contorted, and finally, several different pendulous or weeping types, with their never failing attractiveness. Who said willows are a drab old lot? Incidentally, plant the males every time for maximum beauty and effect, for the anthers are largely responsible for the beauty.

There are upwards of a 100 different types of willows possessing ornamental features and worthy of inclusion in any garden or park—private or public. The fact astounds me that when passing through the countryside or when visiting public parks and gardens one scarcely ever sees an exotic willow, except, of course, the weeping type, and often this is occupying an ill-chosen site and thereby detracting from its beauty. The only others that one might possibly see are the Golden Willow, *S.alba* L. var. *S.vitellina* L. and the Red Willow, *S.alba* var. *S.cardinalis;* also the purple willow, *S.purpurea* L. for its attractive stem colour, *S.daphnoides* Vill. with its lovely catkins in the early spring, and *S.rosmarinifolia*

Wimm. & Grab. for its graceful foliage. There are many more and they will be found in the following list of exotic and ornamental trees and large shrubs:

The caspian or pointed leaved willow *S.acutifolia* Dahl., is a tall shrub or tree up to 7·6cm (25ft) with leaves five times as long as wide; they are elegant, dark green and shiny above, pale grey-green beneath. Young shoots are violet-purple overlaid with white 'bloom'; the stem is upright and also covered with 'bloom'. Catkins appear before the leaves, usually as early as December. Bark, roots and shoots all have a bitter flavour which acts as an effective deterrent to pests.

The Calaf of Persia Willow *S.aegyptiaca medemii* Boiss., is a small shrubby tree from Asia Minor. Its leaves are 7·6–10·2cm (3–4in) long and nearly 5cm (2in) broad. Catkins appear before the leaves in January and February, are very abundant, yellow, very showy and fragrant. In appearance it is beautiful, in usefulness it is an esteemed antipestilential plant, its remedial properties being procured from the catkins by distillation.

The white willow *S.alba* L., is very sparse in Britain but is believed to be much culivated in the north east region of N America, but rarely escaping. It is a large tree of some 18–27m (60–90ft) in height with wide spreading branches rising at 30–50° and twigs which are green and smooth. It is often pollarded and the subsequent growth is used for many purposes by farmers. It has several by-products to its credit and is notable for possessing a large number of relations—sub-species, varieties and hybrids.

S.argentea Sm. (repens var. argentea), has a silvery-grey foliage. It is about 1·2m (4ft) high and the male form bears yellow catkins in great profusion. In its normal form it makes a delightful grey hummock of arching branches but when grown as a standard it is equally effective and closely resembles a white-leaved pear.

S.angustata Pursh., a native of N America in shady woods

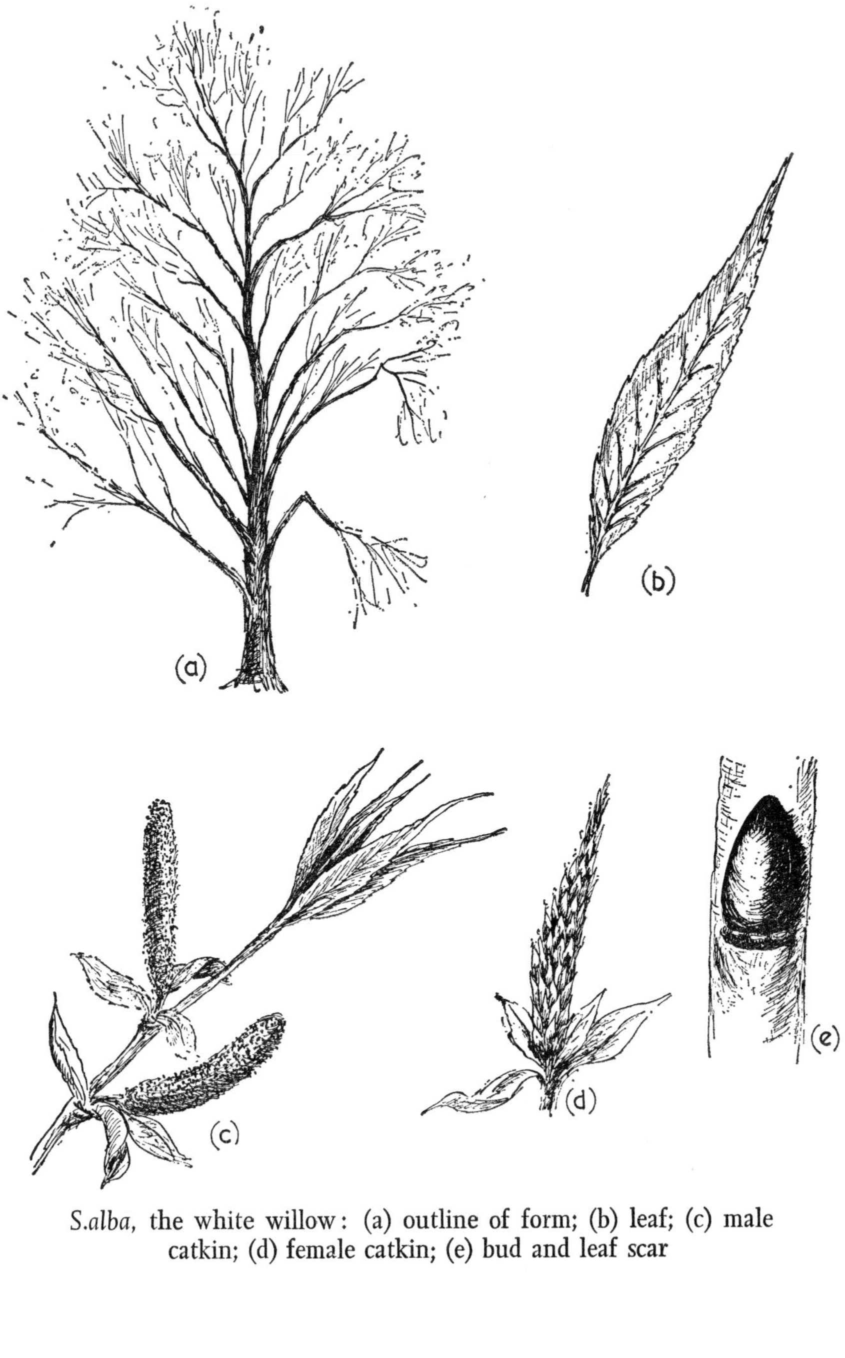

S.alba, the white willow: (a) outline of form; (b) leaf; (c) male catkin; (d) female catkin; (e) bud and leaf scar on banks of rivers in states of New York and Pennsylvania. Very long leaves, tapered and narrow. Flowers March and April. Catkins appear before leaves.

S.*aurea* (*alba* L. var. *aurea*.). Has an upright habit of growth; pale yellow leaves and golden coloured branches.

S.*balfourii* Linton, is a shrubby natural hybrid (*caprea* × *lanata*) having broad grey leaves and long attractive catkins.

The basford willow—S.*basfordiana* Scaling ex Bean (*alba* × *fragilis*), is from the Ardennes and is said by some to be among the most beautiful of willows. A vigorous growing tree with glossy orange-tipped twigs and glabrous tapering leaves, with fine, but sharply serrated, margins. Catkins are elongate and drooping. (Note this feature, for most willow catkins grow erect and not drooping.) A tree ideally suited for parks and open spaces—it is large, impressive and a fast grower.

S.*bebbiana* Sarg. (*rostrata*), the beak willow. Shrub or sometimes tree up to 8m (28ft). Native of high latitudes in North America including Alaska through Canada to Idaho and Montana along the Rockies to Colorado, New Jersey and Pennsylvania westwards to Nebraska. Introduced to Britain 1889.

The two-colour willow S.*bicolor* is believed to be a form of S.*phylicifolia* L., and is a bushy spreading shrub with short yellow branches, villous when young but when older they are yellowish-green and quite glabrous. It rises to a height of 2·4–3m (8–10ft) and has bright yellow catkins. When very young the leaves are purplish with midrib and footstalk, yellow and glabrous. It flowers in April and again in July.

S.*bockii* Seemen, is a shrub from west China and rises to 3m (10ft) in height. Its leaves are 6–15cm (2½–6in) long, deep green above, bluish-white and silky beneath. It is remarkable for its catkins which appear in late summer and autumn and are very crowded and conspicuous. Is very attractive but does not take up too much space.

S.*bonplandiana*, native of Mexico. Introduced to Britain

1829. Stem erect; branches round, smooth and even. Catkins appear after foliage.

S.borreriana Sm. is a form of *phylicifolia* L., a much branched shrub 3m (10ft) high and has large branches which are ash coloured. Twigs are of a deep mahogany hue and glabrous. Leaves are dark green and shining and the buds are large. Catkins are produced in April—earlier than most other mountain willows. When trained on a single stem it forms a very handsome small tree for the suburban garden.

S.candida Fluegg. ex Willd., the sage willow. Is a handsome tree with its ornamental white leaves resembling those of the Black Poplar and flowers early in February or March. Its leaves are snow-white and cottony beneath with prominent midribs. Native of N America, inhabiting cold, damp regions from Newfoundland, Athabaska, south to the United States.

S.caudata (variety of lasiandra). Leaves green on both surfaces and long pointed at tip. Found in Oregon, Washington, Idaho, Wyoming, Colorado, Utah and Nevada.

S.chrysanthos Vahl. (In Oeder), is the golden-flowered Norway willow which takes its name from the hairs on the scales of the catkins; its leaves are glabrous to a great degree and on these features it is quite pleasant but it has no outstanding characters.

S.chermesina (britzensis) Hartig. (*alba* var. *chermesina*), a tree of 6m (20ft) in height. Very attractive in late winter and spring when its young shoots are brilliant orange-scarlet. It is a tree best suited to the park or meadow rather than the garden.

S.chrysostella is a cultivar of *alba* L. var. *vitellina* (L.) Stokes and takes the form of a tree rising up to 4–6m (15–20ft). The brilliant sealing-wax red of the bark makes it a highly decorative plant in winter. Hard pruning in the spring leads to the best effects. When planting this tree two-thirds of the cutting should be below the surface of the soil and it should be planted firmly. Not all willows are so fussy.[9]

S.cinerea L. (ssp *oleifolia* Macreight), a comparatively small bush bearing variegated leaves of yellow on white or yellow on white and red. The twigs and buds are blackish and hairy which tend to make the leaf colouring more striking.

S.coerulea Sm. (*alba* var. *coerulea*) is, for me, the most beautiful of all the upright forms of willows, particularly when it is growing freely, unrestricted and in good cultural conditions. As an amenity tree it will reach a height of 40–45m (120–135ft) in about 15 years and remain beautiful for another 25–30 years, Pyramidal in shape its branches rise at an acute angle of about 30°. With a smoother bark than most, its leaves differ from others in that they lose most of the silky hairs from the under surface early in the season. This tends to give the leaves a bluish-green or bluish-grey appearance—from whence it derives its specific name—coerulea or heavenly blue. The leaves are also translucent and when viewed against the light with a lens the tertiary venation is always plainly visible. This information relates to the pistillate or female tree, which is truly beautiful, easy to cultivate, fast growing and is at its best when sited near to running water. Even the trunk of this tree has beauty in its bark which shows an intricate network of ribs and furrows. It is the prime commercial timber willow tree in Britain. Found wild throughout the north east region of the USA.

S.conformis Forbes., native of N America. Upright mode of growth. Long handsome leaves. One of the earliest flowering of the species—female catkins appear in February.

S.conifera Wrang., cone bearing willow. Wild in N America in shady woods, on gravelly, dry soil, from New York to Carolina.

S.cordata Muhl. (*adenophylla*), is the heart-leaved willow, a native of N America from New Brunswick to British Columbia. It is an attractive shrub about 1·8m (6ft) high with green glabrous branches, red towards the end; leaves are long, smooth and heart-shaped at base. Catkins accompany leaves in March. Flowers very woolly.[9]

S.cramacile Kukome, is known as the mulberry willow. Its special feature is its black catkins which appear in late February. It is related to *gracilistyla Melanostachya* Kukome.

S.crispa Forbes (related to *appendiculata*) is a low growing shrub with pale green branches. Its catkins are small and appear before the leaves in March and again in August. Its anthers are red before they burst, afterwards they turn yellow; its leaves are crisped and spirally curled.

S.daphnoides Vill., the violet or daphne-like willow is a shrub or tree rising to 10m (32ft). Branches are dark greyish, ascend obliquely and are covered with a violet 'bloom'. Leaves are 5–10cm (2–4in) long and 2½–8 times as long as broad; they are dark green and shining above and glaucous beneath. The yellow catkins appear before the leaves from large crimson buds which are dense and fluffy. The buds of this species are unique among willows, in fact, the whole tree is distinctly ornamental and a poor, strong class of soil will meet its needs.

S.daphnoides aglaia, a small tree grown mainly for the beauty of the large, oval, pearly-grey catkins which stud the mahogany growths like strings of beads. Most effective when several of the sort are grouped or massed together.

S.dasyclados Wimm., is a medium-sized tree, sometimes taking the form of a tall shrub. Related to *viminalis* L., its branchlets are densely woolly and its leaves are large—15·24cm (6in) long, lance shaped and woolly beneath.

S.ehrhartiana (*alba* × *pentandra*), is a handsome medium-sized tree with oblong, slender, pointed leaves, usually silky at first, green beneath. Has many of the attractive qualities of *pentandra.*

S.erdingeri Kern (*daphnoides* × *caprea*), a small shrub with silky leaves when young; purple-brown 'bloomy' shoots and attractive female catkins.

Farge's willow—*S.fargesii* Burkill, is a striking shrub of

spreading habit. Young shoots are stout and shining brown; winter buds are large and bright red; it has most elegant leaves—magnolia-like, large, elliptic and glossy dark green. Will add dignity to any garden but requires a reasonable area of ground for expansive growth.

The crack or brittle willow *S.fragilis* L., is a graceful tree of up to 15m (50ft) in height. It has a rounded crown, umbrella shape, with upward spreading branches; the stout bole has roughly-fissured bark and the brittle twigs are yellow-green with a tinge of red. A slight pressure of the finger at the base of the twig will separate it from the branchlet with a cracking sound—hence its popular name. *S.fragilis* is recorded as having been introduced into New England before the revolutionary war, for the production of osiers. It

S.fragilis, sprays with catkins from male and female trees

is widely distributed from Newfoundland to Quebec and southward to Kentucky; Great Britain and Ireland, Sweden, north Russia southwards; west Siberia to Persia and Armenia. One or two of this variety planted in the water meadows, a little way back from the river brink would be quite effective when grown, providing they were allowed to grow freely and not be subjected to pollarding.

S.glandulosea setsuka, a delightful small shrub from Japan. Its chief fascination is in its shoots which are both contorted and fasciated in a most unusual manner; warrants a prominent position in the small shrub section of the garden. Ideal for floral decorations.

The Japanese pussy willow—*S.gracilistyla* Miquel., is a shrubby plant about 1·8m (6ft) high. Its yellow-grey and red catkins are large and very conspicuous in early spring; shoots covered with woolly grey hairs, quite attractive. Popular in E and N America.

S.humboltiana Willd., native of Peru. Cultivated in various places in S America. Introduced to Britain 1823. No flowers produced until 1829. A somewhat tender tree—most suited to the greenhouse for protection.

S.humilis Marsh, the prairie willow. A vigorous shrub of up to 2·5m (8ft) with densely hairy branchlets; leaves 5–10cm (2–4in) long. Male catkins brick-red later turning yellow; female yellow with brick-red stigmas; anthers red. Native to Newfoundland to Minnesota and N Carolina.

The hoary willow—*S.incana* Schrank (*elaeagnos*) is a bush up to 1·5m (5ft) high; branches are long and slender, dark brown, whitish when young, and distinctly warted; leaves are narrow, rosemary-like, 7·6–10cm (3–4in) long, green and villous, covered with white cottony down beneath—young leaves are all snowy white; catkins appear before the leaves. Suitable for the mixed shrubbery.

S.interior Rowe, is the long-leaf or Sandbar willow. A shrub or slender tree. Spreading rapidly with branches from the roots, often forming thickets. Branches erect, smooth with yellowish or purple-red branchlets. Bark smooth dark brown tinged with red. Native eastern Quebec and throughout north east region of N America and from Alabama westward, Wyoming and Colorado.

S.irrorata Anderss., an American upright shrub up to 3m (10ft) with branchlets smooth, purple, rarely yellowish but covered with white bloom. Leaves bright green and lustrous above, whitish- or bluish-green beneath caused by 'bloom'. Introduced to Britain in 1898 from Colorado, Arizona and north Mexico. Very conspicuous in winter and useful in gardens for contrasting with red stemmed varieties.

S.laevigata Bebb, the red willow. A tree up to 15m (50ft) branches orange or dark red in colour, leaves 7·6–17·7cm (3–7in) long, bluish-green and shiny above, whitish beneath. To be found in south Oregon, California, Arizona, Utah and Nevada.

S.lasiandra Bertham, a tree of 18m (60ft) in height. Dark green shining leaves. These are sometimes 6–7in long, ½–1½in wide. Occasionally offered by nurserymen. Native of western N America from the Yukon River, Alaska to southern California, also found in New Mexico and Colorado.

The linear-leaved willow *S.linearis* Forbes., is a low bushy shrub with copious branches. Leaves are 4–6cm (1½–2½in) long and truly linear; they are shiny above and cottony beneath. Buds are bright crimson, branches brown and catkins appear before the leaves. It is easily recognised by its rosemary-like appearance and narrowness of its leaves. Reasonably attractive where competition is not excessive.

S.lingustrina Michx., native of N America, about 4–6m (25ft) high; resembles *nigra* but leaves are longer and narrower and have heart-shaped stipules at their base.

The north American shining willow *S.lucida* Muhl., is a striking shrub or small tree up to 7·5m (25ft) in height. Its lustrous foliage is handsome. Flowers in April and May with catkins 3–8cm (1½in) long. Its one-year branches are greyish-green and smooth, young twigs yellowish-green. Resembles somewhat British native *S.pentandra* L. A very desirable tree for the riverside. Native to N America from Newfoundland to the eastern base of the Rocky Mountains, south to New Jersey and Nebraska, sometimes in cultivation.

The shrub or small tree known as *S.magnifica* Hemsley, is said to be the most remarkable of all willows. Its shape, colour and leaves are all a departure from the willow form, resembling more those of *Arbutus Menziesii* magnolia. In cultivation it reaches 6m (20ft) in height; leaves reach 25·4cm (10in) in length; by 13·9cm (5½in) in width. Young shoots are purple, later changing to red. Male catkins are 10–18cm (4–7in) long but dwarfed by the females at 28cm (11in). The plant was discovered in 1903 in west China at an altitude of 2,743km (9,000ft) and reached Britain in 1909. It is very suitable for large gardens and parks but especially public parks and should be carefully sited to reveal its full beauty.

The Pekin Willow *S.matsudana* Koidzumi, is a tree up to 13m (40ft) high, with spreading branches, yellowish or olive green and glabrous when young. An Asiatic tree with no particularly attractive properties but of a beauty difficult to define. Specimens can be found in several botanic gardens.

S.matsudana tortuosa Rehd., with its branches, twigs and leaves much twisted and contorted is an unusual and curious form capable of adding considerable interest to the garden, shrubbery or edge of ornamental lake. Attains a height of about 4½m (15ft) or slightly more.

S.matsudana umbraculifera Rehd., has a broad umbrella-shaped or semi-globose head. It is a most attractive plant and was introduced from America in 1906.

The hybrid, *S.meyeriana* Rostkov. (*pentandra* × fragilis) is handsome with its grey-white stems, brownish smooth branches, large, broad, smooth, shining leaves and slender catkins. It is a most desirable willow for introducing into ornamental plantations—rather than gardens as it is of the coarser kind. It grows quickly and the male catkin flowers are very ornamental.

S.miyabeana Seemen, is a small, rather handsome tree from Japan, reaching 5m (18ft) in height and with leaves 5–15cm (2–6in) long. The staminate catkins appear before or with the leaves; the anthers are yellow and tomentose. A tree best suited to the mixed shrubbery.

The black willow *S.nigra* Marsh, when grown as a tree reaches 18m (60ft) in height; its branches are irregular, forming an uneven open top; bark is nearly black and heavily furrowed, this unusual feature tends to make it strangely attractive. Found wild in Canada, Nova Scotia, N America from Pennsylvania to Virginia, north Minnesota, east Texas, north Georgia, Florida and Colorado. In the USA and Canada it is considered as of very great importance both for its timber value and beauty of form. It is said to reach a height of 120ft.

S.nigricans Smith is a shrub or small tree up to 4m (13ft) in height. Its popular name—the dark-leaved willow—is apt as the leaves are broad, variable in shape, deep green, rather dull above, paler and pubescent beneath, usually turning blackish when dried. Twigs are pubescent with hairs often conspicuous and white and rather coarse; dull, blackish, brownish or olive green and striate under the bark. Catkins appear with or before the leaves. The female of the species is known as *S.nigrescens* Schl. and is rather rare. I think both male and female are well suited to the mixed shrubbery.

S.occidentalis Rosc., a native of the island of Cuba.

S.octandra Sieb. Sieber deems this to be akin to *S.tetrasperma* Roxb. Wild in Egypt.

The satiny willow *S.pellita* Anderss., is a native shrub of the N American continent and was introduced to Britain in 1918. It rises to 2m (7ft) and its leaves are 5–14cm (2–5in) in length and show a white silky villous effect beneath. The branchlets are 'bloomy'. It is found in Newfoundland to Lake

Winnipeg, south Maine, Vermont, Michigan and also in southern Britain. A handsome shrub in winter with its heavy purple 'bloom'.

The bay willow *S.pentandra* L. (*laurifolia*) is a handsome tree, ideally suited for cultivation. Its leaves are large, smooth, lustrous and dark polished green. When in cultivation it rises to 20m (66ft) in height. Its golden yellow catkins are densely flowered and open rather later than those of any other willow, always after the leaves, usually in late May or June. Compact of form, slow growth and most desirable for the garden, preferably at an end-of-lawn position.

S.phanera Schneid. is a large shrub or tree up to 12m (20ft) in height with very large leaves 6–20cm (2½–8in) long and a native of west China. Quite a handsome tree and merits a place among the ornamental trees.

The phillyrea willow *S.phillyreifolia* Borres., is an attractive, upright, much branched shrub, about 1·5m (5ft) high which flowers about middle of April before leaf-buds open and occasionally a second time in mid-summer. Catkins are 1·27cm (½in) long, numerous and closely set with flowers. Leaves do not exceed 2·54cm (1in) in length, upper surface is bright, shining, full green and bluish beneath. A useful addition to the shrubbery.

In *S.phloragna* we have a willow with one outstanding distinguishing feature; when it has reached tree proportions it annually bursts and throws off its epidermis and outer bark, leaving inner bark, of yellowish-brown colour, completely exposed. This is not beautiful or attractive—just unusual and interesting.

The tea-leaved willow *S.phylicifolia* L., is a shrub of 1–4m (3–13ft) with twigs of polished or burnished chestnut colour at maturity. Leaves are vividly green and shining above and whitish- or bluish-green beneath. A magnificent plant which might well be more widely grown, especially in choice shrubberies.

The Pomeranian willow *S.pomeranica* Willd., is a variety of *daphnoides* Vill. and resembles *praecox gemmata*. Is a very rapid growing plant with branches long, smooth, shining and copiously covered with small yellow dots; preceding year's shoots are covered with a violet-coloured powder or 'bloom'. Leaves are about 10cm (4in) long and nearly 2·5cm (1in) broad, tapering towards both extremities. Catkins appear before the leaves in February and March.

S.praecox gemmata is also a variety of *daphnoides* Vill., but distinguished from it by its very large sized buds containing catkins in the autumn. Both this willow and *pomeranica* are worthy of a place in the small shrubbery.

S.prinoides Pursh., or the prinos-like willow, is a good looking middle-sized tree with oval-oblong leaves and catkins bearing long, soft and curly hairs; these appear before the leaves in March and April. Not outstanding in appearance but would usefully fill up an awkward gap in the shrubbery. Native in N America on banks of rivers from Pennsylvania to Virginia.

The protea-leaved willow *S.proteafolia* Schl., is an upright-growing shrub or low tree exceeding 3·66m (12ft) in height. Branches are of a fuscous brownish-green colour, downy, ultimately becoming smooth; young twigs yellow-purple, covered with hairs. Flowers April and May, catkins 2·5cm (1in) long, thick and obtuse. Quite an ornamental plant for an odd corner of the garden.

The purple willow *S.purpurea* L., is a shrub up to 3m (10ft) in height with slender, tough branches, purplish at first; later grey or olive-grey and smooth. Leaves are up to 3–12cm (1½–4in) long and 3–15 times as long as broad, that is a great variability in width of leaf. Catkins appear before the leaves. By virtue of the elegant slenderness of its twigs during winter, the redness of its catkins, the anthers which are also of that colour before they burst, and the fine purplish hue of its young shoots and leaves, the species is well adapted for planting in ornamental shrubberies. Moreover, the bark and leaves of this plant are so extremely bitter that animals will not touch either—a natural form of protection in fact. Introduced to United States of America as an osier producing species and ornamental willow.

S.purpurea Eugenei., a small tree of pyramidal outline, erect branching; profuse catkin bearer. A very distinct variety and worthy of a place in the garden.

The balsam willow *S.pyrifolia* Anderss. (*balsamifera*), is a much-branched shrub up to 7m (11–12ft) with leaves 4–9cm (1¾–3½in) long; catkins appear with the leaves and are rather loose. It is a handsome shrub, conspicuous in winter with its lustrous branches, shining red buds and red-brown shoots. Most suitable for the small shrubbery. Found in bogs and lowland thickets from Canada to northern New England and westward to Minnesota and south Dakota.

S.rehderiana Schneid., a shrub or tree from west China, up to 9m (30ft) with lance shaped leaves, grey-white beneath. The male catkins have purple anthers. Quite attractive tree for planting on the river bank in a nearby meadow.

S.rigida Muhl., native of N America—New England to Virginia. In swamps and hedges. Very tough. Used for basket making in America.

S.rubens Schrank., a hybrid of *fragilis* × *alba*, intermediate between the parents and rather variable but still attractive with its leaves usually silky when young, becoming smooth, usually whitish- or bluish-green 'bloom' beneath.

The Bedford willow *S.Russelliana* Koch., is a variety of *fragilis* L., and rises to 27m (90ft) in height. Similar to *fragilis* in habit but with brighter olive-brown twigs and coarsely serrated leaves. A recommended tree for parklands and open spaces.

The sage-leaved willow *S.salviaefolia* Linck., is a pleasant form of tree up to 3·6m (12ft) in height. Branches are dark brown, downy when young; leaves are oblong-lance shaped, densely covered with soft, short hairs, whitish or greyish-white. This plant is sometimes used as a stock on which to graft *vitellina* L., with good effect.

S.*sericea* Gaud. (*alba* L., var. *sericea* Gaud) (*argentea*) (*regalis*). Up to 15m (50ft) in height; round headed with silvery white leaves. Distinct from all others in that its catkins remain on the tree for the greater part of the summer.

S.*Smithiana* Willd., a hybrid of *caprea* × *viminalis*, is a tall shrub up to 6m (20ft) in height. Branches are erect, wand-like, round, long, slender, reddish, leafy, smooth, finely downy and soft when young and brittle. Leaves delicately soft to the touch with minute silky down, upper side green, underside whitish. Catkins appear before leaves, numerous and small. Most attractive shrub for large garden cultivation. Not excessively fond of water!

S.*spaethii* closely resembles *stipularis* Sm., but differs slightly in having stouter shoots covered with short, soft, brown hairs and its catkins, which appear in March, are very large and attractive. (Refer to *macrostipulacea* Forbes, in Sallows list for further details).

S.*stipularis* Sm. (believed related *viminalis* L.), a large shrub with broader leaves than viminalis, 12–20cm (5–8in.) long, green, even and soft, downy above, finely downy and whitish beneath, nearly smooth, reddish midrib. Catkins about 3cm (1in) long. Flowers in March. Very rapid grower.

S.*tetrasperma* Roxb., the four-seeded willow, is a native of India where it forms a medium-sized tree, with an erect but short trunk, as thick as a man's body, bearing a very large branching head with twiggy branches; leaves are smooth and have a whitish-green 'bloomy' effect. Flowers appear after the leaves. A rather unusual tree but not in the front-rank of beauty. Of interest in a background situation.

S.*triandra* L., is a shrub, rarely a small tree, 4–10m (13–32ft) high. Bark is smooth and peels off in patches; twigs and buds are brown and smooth. Leaves variable 5–10cm

(2–4in) long, smooth, dark green and shining above, pale green beneath. Catkins appear with leaves. (See also Osier Willow list.)

S.tristis Aiton., is the sad or narrow-leaved willow. Native of N America and found in dry sandy woods from New Jersey to Carolina. Leaves are straight to lance shaped, entire, pointed at each end, smooth above, rugged with veins and downy beneath. Catkins are oblong and appear before the leaves. It has no outstanding beauty characteristics but it should be found a place in the shrubbery as it is exceptionally easy in cultivation.

The Villar's willow *S.Villarsiana* Rouy., a sub-species of *triandra* L. It is a handsome tree up to 4·2m (14ft) in height with dark violet coloured shining branches; the young twigs are dark brown above and paler beneath, polished, angular or striated and very brittle. Leaves are oval in shape, rounded at the base, pointed at the tip, serrated and whitish 'bloom' beneath. Catkins appear with the leaves and the flowers are triandrous. Worthy of a prominent but suitable position in any garden.

S.vitellina Arcangelli (*alba* var. *vitellina*), is a delightful tree and commonly known as the yolk-of-egg willow; its thin twigs are a bright yellow or, sometimes orange. The branches too are bright yellow and distinguish it from other willows. It is particularly effective during the winter season when planted among evergreens. In the shrub area of a large garden the most wonderful contrasting effect can be gained by planting masses of this variety with quantities of white bark honeysuckle (Lornicera xylosteum) the red-barked dogwood (*Cornus alba*) and the brown-barked spiraea (*Spiraea opulifolia*). They are all easy to grow and afford delight in moist situations. Found throughout the north east region of the USA, generally over Britain and many parts of Europe.

S.vitellina Stokes (*alba* var. *vitellina*), is related to vitellina, and produces an equally decorative bark but, in this instance, it is orange-red. When these two related willows are planted in juxtaposition the general effect is greatly enhanced.

S.wehrhahnii, a shrub of 0·9–1·2m (3–4ft) in height, producing masses of silver-white catkins in the spring. These give the effect of the shrub being covered with icicles. It is altogether very attractive, uncommon and worthy of a prominent place in the garden.

S.Willdenoviana Forbes, is a low-growing shrub with green, hairy branches, turning brownish later. Catkins appear in April and again in August. A rather distinct but small and handsome willow—suitable for a front position in the shrubbery.

A shrub of medium height, *S.wimmeriana* Gren. and Godr., is a hybrid of *caprea* × *purpurea*. Young branchlets are grey and hairy, later becoming smooth and lustrous; leaves are silky at first, becoming smooth, dark green above and greyish beneath. A very attractive smallish plant for the shrubbery.

WEEPING OR PENDULOUS VARIETIES

Willow trees are almost invariably unisexual or dioecious, that is, the male and female flowers are carried on separate plants. Occasionally, however, a departure from this rule does occur on the pendulous types—flowers of both sexes being borne on the same trees. The female flowers appear normally on the pendulous or female branches, while the male catkins are found on branches showing a tendency to an upright mode of growth—a sex-distinguishing character. In common with other varieties of willow the pendulous types strike freely by cuttings and grow with great rapidity in a rich soil, within reach of water and in most climates. The shoots are brittle and neither they nor the wood of their stems or branches are ever applied to any commercial purpose.

Up to the present time weeping varieties have been the most popular of all garden willows. In fact, it can be said that the variety *S.babylonica* L., has held a position of honour in the field of ornamental trees over an aeon of time and over a large part of the world where willows are found, including N and S America, Europe and Asia. The day arrives, however, when most leaders must give way to others and this, *babylonica* is now doing, in favour of *S.alba* var. '*tristis*'. But before 'babylonica' finally passes, a few notes concerning it may perhaps not be out of place.

Like so many other members of the genus it is not entirely devoid of confusions and complications. For example, it is not clear as to how and when it was introduced to Britain and why Linnaeus should have named it '*babylonica*' when, without doubt, China is its true country of origin. According to Hortus Kewensis the year in which the first weeping willow reached these shores is shown as 1692, but no further particulars are given. While in exile on St Helena, Napoleon is said to have been greatly attracted to a particular weeping willow and asked to be allowed to sit under it for a time each day. The request was allowed and in view of this rather special use it became fashionable in some parts to possess a plant taken from Napoleon's willow. In consequence a great number of cuttings were removed and distributed widely over Europe, Asia and America. There seems little doubt that many weeping willows flourishing in those continents today could trace their ancestry back to that particular tree.

The beauty of a weeping willow is enhanced when it appears in a scene in which other trees nearby, tend to harmonise with it by their form. This is done partly by the use of the same form of tree, for example, the weeping birch, or partly by contrasting forms, such as the Lombardy poplar. The light and airy spray of the poplar rises perpendicularly, that of the weeping willow is pendent. The shape of the willow leaf is conformable to the pensile character of the tree and its spray, which is lighter than that of the poplar and is more easily put in motion by a breath of air. The weeping willow, however, is not adapted to sublime subjects. It is not suited, for instance to screening broken buttresses, or Gothic windows of an abbey, or the battlements of a ruined castle. The dignity and strength of the oak can support them and is far more appropriate to such tasks. Humbler and more romantic settings, such as a footbridge spanning a gentle flowing stream, or the brink of a shining pond; these are much more suited to the pendulous willow.

For many years *S.babylonica* L., has been afforded a place among the more important shrubs and trees of the well-established nurserymens' catalogues. Much inbreeding between it and other pendulous and semi-pendulous sorts has taken place however, with the result that a hybrid has now arisen possessing much more attractive features than its predecessor. The following extract entry from a catalogue serves to emphasise the amount of inbreeding undertaken with the best species for the purpose of improving a plant which has served its purpose so well and for so long.

> *S.alba 'Tristis'* Gaud. (chrysocoma) (vitellina pendula) (babylonica 'Ramulis Aureis'). The most beautiful large weeping tree in our climate. Awarded the Royal Horticultural Society (London) Award of Garden Merit in 1931.

'Tristis' is indeed a most beautiful tree, hardy in most climates, producing vigorous arching branches terminating in perpendicular branchlets, ultimately of great length and bright yellow in winter. It is fast replacing *S.babylonica* L., but in the large gardens of a few willow enthusiasts the two weepers may be found planted strategically at opposing ends of the garden. It could usefully be sited alongside a stream or lake but this choice is not so much to meet its water requirements which, incidentally, are quite modest, but rather to add to its ornamental attraction by the water-reflection of its long whip-like branchlets which eventually reach the ground. When pond or stream are not available the lawn becomes the

next most usual choice. When this is away from water and long dry periods persist it will be a kindness to the tree if it can be given water occasionally. Before making a final choice of site it would be well to bear in mind that the tree will eventually reach a height of 40–50ft, or even more, with an approximately similar measurement for the spread of its branches. It then could become a truly magnificent tree, but only when it has been afforded adequate space in which to reveal its beauty to the full; this is lost when the tree is cramped for space.

Among other pendulous willows are the following;

S.vitellina pendula Rehd. (*alba* L. *vitellina pendula*). In addition to its weeping propensities this variety has most attractive bright yellow bark.

S.matsudana pendula Schneid., is sometimes referred to as *pekinensis*, the Pekin willow. Its branches are yellowish or olive green and smooth when young. It is an extremely graceful Asiatic tree.

S.purpurea pendula Dipp., is an American weeping tree, with attractive purple stems and very well suited to occupying a damp situation in the small garden, in fact, it has been called 'the poor man's weeper'. It can, of course, look very attractive when placed with others in a willow bed in a large garden.

The hybrid *S.sepulcralis* Simonkai (*alba* × *babylonica*) has similar characters to *babylonica* L., but is of slightly more vigorous growth and less pendulous. One of the best hybrid weeping willows.

Another American hybrid is *S.blanda* Anderss., known as the Wisconsin weeping willow. It is truly a handsome tree having a widespreading head and long pendulous branchlets. Its leaves are a dark green above, bluish-green beneath and very striking. A hybrid of *fragilis* × *babylonica*.

A hybrid between *alba* and *babylonica* is *S.salamonii* Carriere, one of the handsomest and most vigorous of all

S. VITELLINA PENDULA
Outline of form with leaf spray

willows. It is not so weeping as *babylonica* but is still extremely graceful. A female tree, flowering in April, and retaining most of its leaves until December. It deserves to be planted extensively especially in localities too cold for *babylonica.*

The Thurlow weeping willow *S.elegantissima* Koch, is an ornamental tree with long pendulous branches, its leaves are 8–15cm (3–6in) long, bright green above, bluish-green beneath. It is a hybrid of *fragilis* × *babylonica* and is most suited to the medium-sized garden in the urban area.

Strangely enough a pendulous variety exists among the Sallows, which, in that form, is not outstandingly beautiful, but, if planted in the Sallow field, of which I have written elsewhere, it would add gaiety to that place and assume a designation of 'outstanding' among its more modest colleagues. It is *S.caprea* forma *pendula* Th. Lang, and is commonly referred to as the Kilmarnock willow. Its branches are stiffly pendulous.

Finally, there is *S.zabelii pendula,* Vill., a name given to *S.caesia* by Villars for use when the plant is grafted on to a standard and grown as a pendulous variety or form. When so grown it is quite effective in the small garden. (qv page 65)

Whether gracing the lawns and gardens of the great mansions or adding charm and dignity to the humble cottage garden, the pendulous willows are truly delightful and scarcely ever fail to please.

(Pendulous varieties are subject to scab and canker and insect pests. See Chapter 8 for appropriate treatments.)

DWARF SHRUBS AND TRAILING PLANTS

The willows of this group are small of stature and hardier than many other members of a mostly hardy genus. They include the 'little folk' of the willow world—the delightful arctic and alpine dwarfs and trailers. It is strange that the willow should occupy such an insignificant place in most nurserymens' catalogues, particularly when one considers the two species—*S.lanata,* the woolly willow, and *S.lapponum,* the Lapland variety. These two must be among the most beautiful of herbaceous plants—so worthy of a place in any garden. When sited in close proximity to each other their juxtaposition heightens the beauty of each, resulting in a truly delightful effect. When describing the flora of the mica-sehist mountains of Scotland, particularly of Glen Fee, a writer has recorded:

> 'the most obvious floristic feature of a bluff of rock between two long gullies is its display of mountain willows, among which can be numbered, not only *S.lanata,* which is its chief glory, but also *S.herbacea, S.arbuscula, S.myrsinites, S.reticulata, S.lapponum,* and a baffling series of hybrids.'[13]

This high degree of herbaceous beauty is all the more wonderful when one considers the nature of the terrain in which most of these plants have made their native habitats. They provide wonderful examples of associations of sub-tropical vegetation flourishing in glacial conditions—a departure from the generally accepted belief that most species of a given genus are confined to areas within a certain range of temperature.

A group of plants, including numbers of the willow genus, is on record as covering the ground sloping to the beach of Englishman's Harbour in Disco Island (Greenland: lat 69° to 70° N) affording another example of comparatively rich herbaceous vegetation growing on the edge of ice-covered land.[15] The plants are said to be growing particularly well, especially *S.herbacea* L. which is known as 'the least of all' willows. This is a native alpine shrub with leaves seldom exceeding 1·9cm (¾in) in length, that is usually about one and a half times their width but an occasional variation has the effect of making them broader than they are long; they are bright green and shining with veins rather prominent on both sides. The catkins appear after the leaves and those of the male are minute, rarely more than 0·6cm (¼in) long; the female is larger, double the size, in fact, being up to 1·3cm (½in) long.

Its normal habitat is the higher mountain tops and rock ledges up to 1,310m (4,300ft). It is found in many districts of Britain, arctic Europe, and arctic America southwards to the mountains of New Hampshire. It has been described as 'quaint, hardy, surviving extreme cold and exposure and thriving on little more than water and crushed rock'.[11] Its popular name is most apt for not only is it the 'least of all willows' but it is the smallest of all true shrubs. In its wild state its stems divide and creep below the surface rising scarcely more than 3cm (1½in) above it. In cultivation, however, it can acquire a prostrate woody stem of 61–91cm (2–3ft) long and as thick as a little finger. With its branches spread over the uneven soil and often covered with earth, which the heavy rains wash over them, they present the phenomenon of being more or less subterranean.[13]

A place where the species can be found growing in great profusion is around Lake Louise at the foot of Saddleback 2·44km (9,000ft) in the Canadian Rockies—the land of glaciers and everlasting snows. By the nicety of its choice of situation it ensures an adequacy of water for its needs and in the few brief hours when the sun casts its warming rays over the 'willow forest', bees take advantage of this warmth and draw on the banquet of nectar provided so lavishly by 'the smallest trees in the world'.

Incidentally, *S.herbacea* L., is a perfect tree in a dwarf state, not a man-produced miniature. When planted on the rockery it will trail over the rocks with gay abandon, or it may be placed in a rectangular earthenware container, its shallow soil being decently covered by a thin layer of small pebbles. It could then be allotted a place on the rockery with equally good effect. This form of treatment is also applicable to a number of the other dwarf willows.

The following are some additional dwarf and trailing varieties:

S.apoda. This is a dwarf alpine with prostrate or low-spreading branches and is most picturesque in its mode of growth. The female has long grey catkins in the spring while those of the male are pink and orange coloured.

S.arbuscula L. In its wild state this willow flourishes on damp ledges of basic rocks, so for those persons with rock gardens it would be an act of kindness to *arbuscula* to place it among the larger stones of the garden rockery—it would be quite at home there.

S.arctica Pallas, is a creeping plant with thick branches and catkins 3cm (1¼in) long. It is the most northerly woody plant known and is a native of the northerly parts of America—Hudson Bay and Melville Island. It will make an attractive addition to the rockery.

S.arenaria L., the sand willow, is for those people who are not green-fingered, the ideal willow to plant. Not that the willow genus is difficult for it is not, but *arenaria* is perhaps the most responsive of them all. Its roots are many yards long even when the plant is still small. In its native Scotland it flowers in June; in slightly warmer climates the flowers arrive in May.

S.betulifolia Forster, is a sturdy upright bushy shrub of about 60cm (1–2ft) in height, with abundance of short, leafy, dark purplish branches, hairy when young and not downy. Leaves are rigid and thin, of crackling and veiny texture. Not one particularly appropriate for the rockery but rather more at home in the small-plant shrubbery.

S.boydii Linton, is a hybrid of *lapponum* × *reticulata* and takes the form of an erect shrub, sometimes forming a miniature gnarled 'tree' of about 30·5cm (1ft). Its leaf blades are very small, up to 2cm (¾in) long, rotund and silvery grey;

catkins are up to 2cm (¾in) in length and appear with, or rather before, the leaves. It is a delightful plant and ideal for the rockery.

The alpine grey-leaved willow *S.caesia* Vill. is a low, straggling shrub 90–120cm (3–4ft) high. Leaves are about 2·5cm (1in) long, with short sharp points. Catkins are 1·2–2·5cm (½–1in) long. (When grafted on to a standard and grown as a small pendulous plant it takes the name of *zabelii pendula,* Vill.) (qv page 61)

The cotinus or quince-leaved willow, *S.cotinifolia* Sm., is found in woods, on banks of rivers and is normally about 61cm (2ft) in height. When brought into cultivation and planted in a sheltered position it has reached 2·4m (8ft) in height. It is always upright in its method of growth, has straight, round, brown, downy, moderately spreading branches and its catkins appear very much earlier than its foliage.

An attractive little shrub, *S.decumbens* Forbes, is a variety of *repens,* and has downy branches extending obliquely from the ground up to 30–45cm (1–1½ft). Leaves are dull green and lightly silky above, densely silky beneath. Buds are red before expansion and the catkins are nearly 2·5cm (1in) long. Easily grown and is not at all fussy about type of soil.

S.divaricata Pall. In its native Dauria it grows among granite rocks of the Alps and spreads in prostrate form. The leaves are crowded about the ends of the shoots and both sides are quite smooth. Another gem for the rockery.

S. gillotti G. and A. Camus, is a natural hybrid of *lapponum* × *phylicifolia* and from that breeding one can expect only good results. It is a small but wide-spreading and semi-prostrate ground-cover plant—entirely suitable for large rock gardens.[5]

S.glauca L., is the glaucous mountain willow, stout and bushy, up to 91cm (3ft). Leaves are nearly entire, about 5cm (2in) long, a beautiful glaucous green above, woolly and snowy-white cottony surface beneath, slightly prominent veins and a reddish midrib. In the Highlands of Scotland it flowers in July but in warmer climes quite a month earlier. The description sounds attractive and that is no departure from reality. It is native to N America and Canada, Alaska to Yukon, and Alberta and Greenland.

S.grahamii Borres, is a true Scot, a native of Sutherlandshire, and a hybrid of *herbacea* × *myrsinites*. It takes the form of a small shrub, growing about 30cm (1ft) high.

S.hastata L., is a small spreading tree of up to 1·50m (5ft) in height. As it is found usually in moist places it would be sited best at the side of a pond or stream. When discovered it was growing by a small stream that passes through the sands of Barrie, near Dundee.

S.herbacea L. 'The least of all willows' is an alpine shrub with aerial twigs usually not exceeding 3cm (1½in) with 2–5 leaves on each; long creeping branches and underground stems. (This species has been more fully dealt with earlier in this chapter.)

S.kitaibeliana Willd., a very small shrub with yellowish glabrous branches spreading close along the ground. In the willow garden at Woburn Abbey, Bedfordshire, it flowered in April and May and again in August.

S.lanata L., is a bush of 60–120cm (2–4ft) in height and known as the woolly willow. Its entire leaves are woolly-silky-tomentose when young, finally dull green and glabrescent, but persistently woolly-silky-tomentose beneath, sub-glaucous and prominently reticulate-veined. In Sweden, it is considered the most beautiful willow in that country, if not in the whole world. The splendid golden catkins at the ends of the young shoots light up, as it were, the whole bush, and are accompanied by the young foliage sparkling with gold and silver. Also its yield of honey is greater than that of any other willow which points to the wisdom of siting it near the hives. If grafted on to a standard it would make a delightful spring flowering tree, particularly suited for the larger suburban garden.

S.herbacea, branches, leaves and catkins

S.lapponum L., is the Lapland willow, a shrub of 30–120cm (1–4ft) of compact habit, usually erect, but occasionally trailing. Leaves are small and crowded at ends of twigs. Quite a beautiful shrub and worthy of a space in every garden.[18]

S.malifolia Sm., the apple-leaved willow is a spreading shrub of 90–180cm (3–6ft) in height; twigs are crooked and brittle; flowers in April and its foliage resembles that of an apple tree rather than a willow. It is quite attractive.

S.medwedewii Dode., a small shrub, with narrow lance-shaped leaves, 6–10cm (2½–4in) long and 5–6mm (⅕in) wide; dark green on upper surface, bluish-white beneath. A native of Asia Minor.

S.microstachya Turcz. ex Trautv, is a shrub up to 30·5cm (1ft) in height; leaves 1·5–4cm (⅝–1½in) long, hairy and shining when young. Distinct in the extreme narrowness of its leaves; native of the Caucasus. A rather unusual type of willow but interesting.

S.moorei White, a small rather unusual type hybrid with prostrate branches and some erect branchlets. With its polished dark green leaves it is well worthy of a place in the rock garden and would function well as a ground-cover plant.

The American brown willow, *S.muhlenbergiana* Willd., is a small shrub, mostly procumbent, indigenous in gravelly places in Pennsylvania and Canada, also in dry shady woods from New York to Virginia. Branches are greenish-yellow with black spots; anthers are purple, yellow when they burst; bracteas are white, tipped with red. Of extremely pleasing appearance. Does not require wet or even damp situations.

S.myrsinites L., another rock garden prospect. The whortle willow is a small wiry creeping half-erect shrub of 10–45cm (4–18in.) in height. Dead leaves persist until end of following season. Blossoms with great freedom and is ideal for the rock garden.

Page 69 (above) S. herbacea *is a truly tiny plant in its wild habitats but will grow three times that size in cultivation in temperate regions;* (below) S. apoda *is a dwarf Alpine willow, picturesque in its mode of growth; its female catkins are long and grey while the male are pink and orange*

Page 70 (right)
S. boydii is a delightful hybrid in the form of a small gnarled woody tree about 1ft high, arrayed with silvery-grey leaves; (below) The whortle willow, S. myrsinites, is a wiry little creeper which blossoms freely on the rockery

S.myrtilloides L., a small-leaved dwarf bush from about 5–90cm (a few inches to 3ft) high, normally decumbent with a creeping subterranean stem. Its general appearance is well portrayed by its popular names—the myrtillus-like or bilberry-leaved willow.

S.oborata Pursh., a native to Labrador and to the north-western coast of America—a pleasant small willow but not outstanding.

S.obtusifolia Willd., a hybrid of *aurita* × *lapponum*. A slender shrub not exceeding 1·3m (4ft) in height. Native to the woods and mountains of Lapland. Quite a pleasant willow to have around but no outstanding features to mention.

S.petrophila Rydb., is a small-leaved creeping plant from the American arctic regions related to *S.arctica*, rising only about 10cm (4in) above ground. Native of the mountains of N America, Canada, California, Montana, Colorado and north Mexico.

S.polaris Wahl., the polar willow is similar in habit and shape of leaf to *herbacea* L., distinguished mainly by leaves being almost invariably entire and smaller on average.

S.procumbens Forbes. This is a low procumbent plant extending along the surface of the ground with greenish-brown, shortish branches; leaves are bright green and shining on both sides; flowers in May and June in large elongated catkins, thick and cylindrical. It is a truly beautiful shrub and will add to the beauty of any garden.

The willow *S.prostrata* Smith, is worthy of note, even if not outstandingly attractive. The stems comprise an entangled mat, several centimetres in diameter, with straight, slender, round, leafy, tough, downy or silky branches, 30cm (1ft) or more in length, spreading close to the ground in every direction, with a few short upright ones occasionally; catkins are numerous, appearing before the leaves, flowering in May, and in the wild the plant prefers moors, heaths and sandy situations. When bringing this plant into cultivation it is wise to offer it similar cultural conditions, at least, insofar as site and soil are concerned.

S.pyrenaica Gouan. The stems of the pyrenean willow are quite prostrate, branched and smooth; leaves are 2·54cm (1in) long by about 1·27cm (½in) wide, bright green and shining above, woolly about the margin giving a peculiar but characteristic appearance; female catkins 5·08cm (2in) long and slender. Native of the Pyrenees, contiguously to the region of snow. Flowers in May and is quite attractive.

S.radicans Sm., is the rooting-branched willow, a spreading bush with long purplish branches taking root as they extend. Leaves are 5cm (2in) long, less than 2·5cm (1in) broad, very acute at point. This is, of course, a plant which may be utilised on the rockery with good effect because of its free branching features.

The creeping willow *S.repens* L., is a very attractive small-leaved shrub normally prostrate or slightly ascending, 30cm to 2·4m (1–8ft) high. The latter height is rarely attained except in cultivation. Stems prostrate to erect, slender, spreading by creeping rhizomes; leaves glistening and silvery beneath; catkins appear before the leaves in April and May. Damp, rocky heaths preferred when wild, so damp rockeries would appear as suitable for cultivation.

S.reticulata L., the reticulate, netted or wrinkled-leaved willow, is a creeping shrublet seldom more than 17·5cm (7in) high; short few-leaved twigs, short-branched rhizome or creeping and rooting stem quickly becoming an intricate mass of dark glossy underground twigs. Leaves do not exceed 3·8cm (1½in) in length and half those figures in breadth; catkins purplish-red appearing after the leaves in June or July. While it is slightly larger than *herbacea* it either mantles the alpine rocks or spreads on the ground in large patches; is often partially covered with moss. It is often found on damp heaths

S.repens, the creeping willow; leaves and male and female catkins

S.reticulata, leaves and male and female catkins

where its catkins peep up above the heather from a smooth orange stem that roots down at intervals. *Reticulata* is attractive, unusual, singularly unwillow-like in appearance, but unfailingly recognisable, for it resembles no other shrub and varies but little. The kind of soil it obviously prefers has been noted as micaceous, that is, where the crumbled crystals of rocks of silicate aluminium and other silicates are found in small glittering scales in the granite. Your nurseryman will arrange to supply something of this order for you at time of purchase of plant. It is certainly a plant for your willow bed or rockery and may be found in arctic N America from Alaska to Newfoundland, Alberta, Greenland, and south to the Rocky Mountains and Colorado, also in Britain and mountainous areas of west Europe. It is a remarkable species and justifies inclusion in any collection.

S.reticulata var. *prostrata*, an even smaller plant than its species (above) and very much slower in its rate of growth. Quite worthy of a place among the small varieties.

S.retusa L., the retuse-leaved willow is a low prostrate alpine shrub, reaching only a few centimetres above ground, the branches creeping and taking root. Leaves are lozenge-shaped and tiny—0·85–1·9cm ($\frac{1}{2}$–$\frac{3}{4}$in) long, 0·25–63cm ($\frac{1}{10}$–$\frac{1}{4}$in) wide. It will spread freely in the rock garden and forms close tufts in exposed places—in fact, it is a good ground-cover plant. Flowers in May.

S.rosmarinifolia Wimm. and Grab., is the most distinct of the creeping willows. Long linear, oblong leaves and small catkins. In gardens however it can aspire to a tall bush with fine branches but it can be kept in its dwarf state by hard pruning; prefers a moist, sandy or turfy situation and flowers in April and May.

S.sadleri Syme. Here is a distinct hybrid of *lanata* × *herbacea* closely resembling *lanata* in its woolly characters.

S.sericea Gaud (*petiolaris* var. *sericea*.) The silky willow, with a prostrate stem and entire leaf, silky and hoary on both sides. The catkins also are silky and about 1·27cm (½in) long. Quite a pleasant plant but nothing spectacular about it. Native of N America from New Brunswick to Tennessee, inhabiting damp places.

S.serpyllifolia Scopoli, is a very curious little plant, commonly called the wild-thyme-leaved willow. The leaves are ovate entire and shining; catkins are oblong with very few flowers which appear in May. Really this has not very much to offer by comparison, of course, to the many other much more striking varieties.

S.silesiaca Willd., is a sub-alpine plant related to *caprea.* Its leaves are up to 14cm (5½in) long, silky when young, later nearly smooth and pale green. Flowers in May.

S.stuartiana Sm., an attractive little willow, small-leaved and shaggy; leaves nearly entire, densely silky, somewhat cottony beneath. Bushy type, copiously branched, rising to 61–91cm (2–3ft) in height, sometimes rather more. Would look smart at end of ornamental pond.

S.uva-ursi Pursh. Here is a beautiful little species with all the appearance of Arctostaphyles *uva-ursi* in habit, as well as in the form of its leaves. It flowers in April and May and is known as the bear-berry-leaved willow and is certainly worthy of a place in the willow garden. It is native to Labrador, Alaska, N America and Newfoundland.

S.vestita Pursh., is a low, elegant, prostrate shrub up to 20cm (8in) in height, with obovate or elliptic leaves 2–5cm (1–2in) long, silky above at first but soon dark green and glabrous. The catkins appear on very leafy branchlets. Most attractive for the rockery.

S.villosa Forbes, a small shrub, with slender, greenish-yellow, villous branches, sometimes having yellow dots; leaves are finely serrated towards the tip and slightly toothed and tapering towards the base; upper surface is also greenish-yellow and shining. The combination of form and colour of both leaves and branches is very different from any other species of the genus. Here is something different for the small shrubbery.

S.Waldsteiniana Forbes, an alpine variety of no particularly extravagant features; it has nothing to indicate that you should or should not include it on your rockery—but no doubt you will.[18]

4 : *Commercial and other Types*

No PART OF the willow structure is quite without its uses; many are, in fact, quite valuable. The root, trunk, branches, twigs, leaves, bark, flowers and various interior substances—according to the variety of willow—all have some purpose in the interest of man.

TIMBER PRODUCTION

The prime timber commodity of willow is the cricket bat and, in a lesser degree, artificial limbs. The first of these is made exclusively from the coerulean variety, *alba*, var. *coerulea*. Nowadays the majority of artificial limbs are made of light metal alloys, although a quantity are still manufactured from wood, and where this is so the coerulean timber is used. Cricket bat blades are, however, the prime commodity, having been made from this particular variety of willow since about 1820. It is of interest to recall that the game of cricket has been played since the early fourteenth century when the bats were cut by the individual players from willows growing wild along the banks of streams and other such places. The bats, so called, were scarcely more than roughly hewn clubs bearing little or no resemblance to those in use today. In time, players demanded greater uniformity in their bats, particularly in regard to such features as weight, shape, size, colour, toughness and strength, and it became obvious that such requirements could be met only by the use of timber possessing such qualities. Accordingly, about 1780, a search was undertaken by players, foresters, botanists and other interested persons. Several possible timbers were discovered but in no instance was the sex of the tree known. The importance of a knowledge of this factor becomes evident when it is realised that only the female coerulean willow is capable of producing the required standard of timber. The male of the variety is quite worthless and very sparse in its numbers. (I have seen hundreds of thousands of willow trees in the course of my wanderings but only 3 male coerulean trees!)

Success in the search for a suitable timber came in 1803 when Dr. James Crowe of Norwich discovered a willow plant growing wild which he believed was not only of the coerulean variety but also of the female sex. The tree was found in the parish of Eriswell in north-west Suffolk, a district which had, and still has, an extremely large and varied willow population. Crowe removed his 'find' from Suffolk to Norfolk and planted it in his salictum at Norwich. Twenty years later the tree was felled and his earlier opinion about variety and sex were then confirmed. In due course sets cut from the tree were being distributed not only among his friends but more widely among nurserymen in East Anglia and south-eastern counties of England. In less numbers sets were being despatched to the southern, western and north-midland counties; also to the low countries of western Europe and in still smaller numbers to eastern and northern areas of North America. I am not aware that this particular variety is grown in large numbers elsewhere.

The first essential in cricket bat timber production is to ensure that the willow trees employed have their origin in the plant discovered in the nineteenth century in north-west Suffolk. The need for this became clear in the early days of the present century when it was found that no more than a quarter of the trees then being planted were of the true cricket bat variety; twenty-five years later large numbers of those same trees, having reached maturity, were rejected by buyers as unsuitable. An investigation of the problem revealed that the rejection of certain trees was not traceable to the influence of climate, soil or moisture, but rather to the systematy of the tree. In those days sets were taken from pollarded trees, many of which were of doubtful origin. Also, most growers could not tell from the appearance of a set or even of a young tree whether it was true *coerulea* or an inferior hybrid resembling it. For that matter many growers would be in much the same situation today were it not for the knowledge and integrity of reputable suppliers of sets. It is here that real advance has been made over past years, for the production of sets is now largely in the hands of a few reliable merchants and advisers whose basic plants are guaranteed to be of true stock. As the mechants are the main buyers of mature trees it is in their own interest to see that only true bat willows are planted. Nevertheless, it is advisable that the grower himself should have a knowledge of the main botanical features distinguishing the true bat willow from the rest.

The number of bat willows in cultivation at any particular time is difficult to assess due to the absence of official records, also to fluctuations in fellings and plantings. I have taken figures based to some extent on information supplied by the larger growers, but more on my own personal observations in the course of my pathological work on the trees. The figures indicate that at any one time there can scarcely be less than two million trees in cultivation, half of which are in Essex and East Anglia, and the remainder widespread over other parts of Britain. However, these figures need not deter potential growers from entering the industry, for not more than about 5 per cent of the trees produce first grade timber, which has the effect of creating a continual shortage. This situation is due in some measure to neglect on the part of growers to observe the essential cultural procedure in the early stages of growth and also to the ravages of Watermark Disease.

With the comparatively recent introduction of both plastic-covered and laminated wood-faced bats, which can be produced with the basic willow wood being of slightly inferior quality, the prospects for commercial willow growers still appear to be good. On the assumption of the continued large home demand for bat clefts, plus an ever-increasing overseas market, the continued utilisation of all first-class sites for the production of high grade willow timber, seems a commercially sound proposition. The ultimate end of these fastidious trees should justify the toast of Bowen in his immortal Harrow School song: 'Honour and life to Willow the King'.

Coerulean timber, when fast grown, is appreciably lighter than that of most others. At the time of felling, the weight of the 'other sorts' averages 44lbs per cubic foot and, when air-dry, that is with about 15 per cent moisture content, it is approximately 29lbs per cubic foot. By comparison, the coerulean willow averages 35lbs and 22–6lbs respectively, according to the speed of growth of the trees and other factors which might have occurred during the period of cultivation. The sap-wood is off-white and the heart-wood reddish brown, although the latter can vary under certain circumstances. Indeed, the heart-wood of some species of poplar is reddish-brown, closely resembling the willow, and this fact has in the past tended to confuse slightly the recognition of the timbers.

Makers of cricket bats, however, are scarcely ever confused on this point and generally recognise the timber with certainty by its appearance and manner of working. The annual rings are distinct on clean-cut transverse surfaces, becoming less distinct near the centre of the tree. For the best bats, sapwood consistently light in colour and having growth rings of up to one inch in width, are called for by both manufacturers and players. The last formed annual ring, that is the outermost layer of timber, is of inferior quality to the inner sap-wood and it also contains an abundance of starch, which renders the wood liable to attack by *Lyctus* beetles—hence the value of removing it before the clefts are stacked for seasoning.

The particular qualities of cricket bat timber most desired are lightness, resilience, strength or toughness, white colour throughout, silkiness of texture and straightness of grain, all of which must be present for it to be considered as first-grade timber. Cricket bat willow is cut at a younger age than any other of our timber trees. Trees of 10 to 12 years of age are often five feet in girth and big enough for felling. They are usually sold standing to merchants who make a special trade of cricket bat cleft preparation. Chelmsford in Essex and Bungay in Suffolk are the two main centres of the business. Soon after felling the trees are cross-cut with saws into lengths of 2ft 4in, the standard size for blades of bats; three, sometimes four, such lengths are obtained from each trunk or bole. These sections or 'rounds' are then cleft, either in the fields or at the factory, into wedge-shaped segments radiating from the centre. On the outer tangential surface each cleft is about 4½in wide and each 'round' of timber of about 44in, overbark measurement, may yield eight such clefts, that is 24 or 32 according to whether it is a 3 or 4 'round' tree.

WILLOWS FOR OSIER PRODUCTION

The name 'Osier' comes from the French osiere and the Latin 'ausaria' meaning a willow bed. Osiers are the long straight rods which are grown on low willow stools planted in beds or 'holts' mainly in the following willow growing areas: Somerset, Gloucestershire, Nottinghamshire, Berkshire, Suffolk and Lancashire; overseas the Argentine has the largest production figures, leading Germany, the Netherlands and several Dutch possessions, Belgium, Poland, Portugal and a few British overseas possessions, roughly in that order.

In delectable Somerset, in the south-west of England, the mantle of King Alfred (of the burnt cakes fame) is still spread over the Isle of Athelney (Royal Mantle of the Sea, gold clasp of the sun, grey minever of the willow). As the south-western breeze sweeps the grey and yellow pollen, all fine and soft, from the pistillate flowers of the willow, the men of Somerset, strong of heart and hand, plant and harvest, reclaim and plant again. They are producing willow rods for basket manufacture from the several willow species proven especially suitable for such production. Somerset, with about 1,000 acres in production, is among the largest willow areas in Britain. Passengers travelling by rail from London to the Cornish Riviera can catch a glimpse of the main osier beds a few miles before reaching Taunton. They may even see some rods cut, peeled and drying on wire fences along the road sides. Travellers by road who care to spend a little time hereabouts will quickly gain the impression that osier production is based upon 'terms' and find it fascinating, almost attractive and certainly complicated—at least it will seem so to the layman.

The term 'hard rods' is determined by their having a high proportion of wood to pith, plus the character of the wood tissues in the rods. The rods of *S.viminalis* are inclined to be more open grained in their wood than those of other species and are termed 'short rods'; they contain more pith than wood and, in consequence, possess working qualities somewhat inferior to the 'hards'. 'Hard tops' and 'soft tops', 'full tops' and 'fine tops' are among other descriptive terms in use. As part of the preparation of the rods for market there is the 'whitening' carried out by such processes as 'couching', 'pieing' and 'pitting'—the object being to retard the flow of sap and extend the period over which the rods will peel satisfactorily. Where the grower has no facilities for preparing the rods, they are marketed direct from the field as 'green' and sold at appropriately low prices.

'Preparation' of rods includes peeling, thereby turning them into 'whites' or 'buffs'. The people who perform this work are assisted by 'fixed brakes' plus a spirit of cheerfulness that does one's heart good to behold; they are typical of folk who work close to the land. Standing by their 'brakes' which are fixed to tables or posts, they each place a bundle of green rods at their side with all butt ends near the 'brake'. Taking

S.viminalis, leaves and male catkins

each rod separately they draw the butt end through the 'brake', splitting the skin on the rod, then reversing the rod, they again take it through the 'brake' thereby removing the skin and leaving the rod to come clean and white. The 'buff' rods are obtained, first, by preparing from the 'green', then loading into long water tanks to boil for several hours. This is known as 'buffing' and is followed by the peeling process as detailed for the 'whites'. The rods having been stripped of their skins are placed on end in the open air to dry. As the drying continues, colouring develops. The rods that have been neither 'buffed' nor 'whitened' are known as 'browns' and sold with their skins on, at a low price.

Still on the question of terms or names the Berkshire growers' gradings include 'tack', 'short small', 'long small', 'threepenny', 'Middleboro' and 'large'; and from the Mawdesley area of Lancashire we have 'long scanes' and 'red buds'. It is said that one-year-old rods in the Argentine will attain a greater length than those grown elsewhere but more important, they are without a corresponding increase in diameter at the butt; thus they are more analogous to cane and consequently more satisfactory to work with.

From time immemorial willows for osiers have been propagated from cuttings. In fact, the methods employed today are substantially the same as applied in Roman times. Different species of willows produce appropriate rods for differing products. The rods form an easy prey to insect and fungoid pests which damage the leaves, terminal buds, rods and stumps. (See Chapters 7 and 8 for details of treatments.)

In the production of basket willows it is important to determine whether the male or female plant is the more suitable. There can be no doubt that the female of most species is the more vigorous and, consequently, where speed of growth and ultimate strength of product is required the female of the species is preferable. On the other hand when tough, yet delicate, rods are necessary for the finer basket making, not only the finer growing species but the males of the appropriate species, should be chosen. Osiers for hoops are produced on *S.viminalis* and *S.caprea* today, even as they were in Holland, at the epoch of the Dutch learning to pickle their herrings and pack them in barrels, which they were taught to do by Beukelson, who died in 1397. The Dutch boers, without knowing anything of the sexes of willows, selected those plants which appeared to them to be of the most vigorous growth and thus, without knowledge or intention, they propagated only the female willows. The cultivation of osiers in close proximity to basket making factories constitutes a combined industry with good effect. Naturally the sorts grown will be those most suited to the particular business thereabouts. A similar principle is employed in many other localities—Cambridgeshire for potato baskets and Gloucestershire and Kent for fruit baskets. Smaller beds exist in the Devon seaboard coombes for the provision of material required for the making of lobster pots. Fine baskets are woven from willow rods at Abercych in Pembrokeshire. A special type of boat-shaped basket indigenous to the Highlands is produced from willows grown in the Island of Skye.

The most suitable soil for the growing of osier willows is a loam of from 6–8in in depth with a stiff marl or clay sub-soil. However, the land should also allow for irrigation during dry summers with the object of maintaining a constant healthy growth. As a rule, arable land is not equal to that of the meadows as the former usually lacks the high state of fertility which follows the breaking up of old turf. Osier willows do not grow so well on the outside, as on the sheltered inside portion of the field, the former becoming exhausted by tree and hedgegrowth. A good heavy dressing of farm manure well ploughed in will help to put this right. The checking of weed growth and regular manuring are essential factors in the cultivation of this crop.

When the shoots or rods of willows are grown commercially

Page 87 (above) S. pyrenaica *is a native dwarf of the Pyrenees, and very attractive;* (below) S. repens *accompanied by Marram grass on sand dunes on the north-west coast of England*

Page 88 (above) S. reticulata *is a shrublet seldom more than 7in high. An ideal plant for the shallow pot or rockery;* (below) *an even smaller plant than its species is* S. prostrata, *a variety of* reticulata *and a native of the Dolomites*

the shoots should be ripened annually. In cold climates where the plants are grown in soil which is abundantly supplied with water late in the season this is difficult to achieve. Hence, the colder the climate the drier should be the soil, if for no other reason than to meet the necessity of perfectly ripening the wood. Incidentally, it is a well-founded principle that the poorer the soil the more often the wood that grows on it should be cut over.

Somewhere between a dozen and a score of different species of willow are employed in the production of osier rods and, not surprisingly, the quantity and quality of the rods produced varies widely as between the willow species from which they are cut and the districts in which they are grown. The species mostly employed include the following:

The white willow *S.alba* L., produces Africans, the better of two kinds of rods which are largely imported into Britain. It is an excellent osier when produced in warm, humid places.

The rods of the golden willow, *S.alba* L. var. *vitellina*, when used with the bark on, in a green state, are said to be the toughest osiers grown. They are slim and chiefly produced in the Reading district and in Suffolk. They are sold for tie-rods, to market-gardeners, nurserymen and celery growers.

The Belgian Red Willow, *S.alba* L. var. *cardinalis*, produces the best working quality osier that Belgium exports. The product of this willow, like the previous one, is intended for tie-rods, which find a ready market with gardeners and nurserymen.

The almond-leaved willow, *S.amygdalina* Sm., is grown in many districts and produces rods of French origin. Brunette Noire or black, Grisette Droit or straight, and Grisette Courbe or bent, are all of excellent quality. Trustworthy is tall and vigorous and comes from Suffolk and Berkshire. Whissenders is grown extensively in the Trent Valley.

The only sort of rods produced on *S.daphnoides* Vill., the violet willow, are known to the trade as Violets because of

S.amygdalina, leaf spray and female catkins

their violet 'bloom'. They flourish in poor, strong soil and produce abundantly.

A vigorous growing willow and a heavy cropper is *S.hippophaifolia* Thuiller, known as the Sea-buckthorn-leaved willow. It prefers a rich soil and is well adapted to the method of cultivation practised on sewage farms. It produces a mass of dense foliage which effectively keeps the undergrowth in check.

In addition to its qualities as an ornamental tree the Bay willow, *S.pentandra* L. (*laurifolia*) is a most efficient osier-rod producing variety. The names of its rods are Lumley, Patent Lumley and American Green and, incidentally, all are very well suited to transatlantic climatic conditions. The species did, in fact, come to Britain originally from Maryland and Pennsylvania, where it is still grown.

The purple willow *S.purpurea* L., is another quite graceful shrub and used widely for ornamental purposes. Its osier-rods are the most slender, for their length, of all willows. They are easily recognised since all are yellow on the inside of the bark. They are very bitter to the taste and show red eyes at growth. Rabbits and cattle will seldom touch them. It is also grown in America for rods named Dicky Meadows and Red Buds. These rods run along the ground like strawberry runners, are very beautiful, wiry and heavy cropping. Incidentally their form of growth makes it difficult to keep the ground clear of weeds; weeders must work unshod with their feet clothed in some soft fabric to avoid bruising the willow shoots. Two other beautifully slender rods are Pyramidalis from Germany and Brittany Green from France, off the same species.

S.triandra L. is normally found growing as a shrub, rarely as a tree. It is widespread and common in England and many other parts of the world. When grown for the production of osiers it thrives on a cool, strong loam and makes its best growth in a wet season. While its varieties of rods are not legion, the rods are many in number. Among some of them

S.amygdalina, Almond-leaf willow; outline of form

are Black Maul, good for baskets which are subjected to long and arduous service; Black Italian, a superior and still harder rod; Black German ranks high among osier and is a heavy cropper; Lincolnshire Dutch is a vigorous grower and thrives well on heavy warp land; Raynes Ten-feet is usually employed for large white hampers; Black Holland is one of the largest and longest rods and was brought to East Anglia by the Dutch; while Champion Rod is grown mainly in Somerset.

The Common Osier, *S.viminalis* L. Introduced to America for wicker work. Occasional escape from Newfoundland to Pennsylvania. Is a shrub 3–10m (9–30ft), with long, straight and flexible branches. As a producer of osier rods it embraces many varieties differing widely in their working qualities. Shoots on this species will reach 3·66m (12ft) or more in one season. If left on the stools for two or three years they will produce strong, lengthy straight sticks which demand a ready sale to makers of hampers and basket furniture. (Incidentally, I am told that the term 'Osier' is correctly used when applied exclusively to *S.viminalis* L., and its varieties.) In addition to the more normal use for the rods they are suitable also for holding up river banks and for wild-bird coverts in low out-of-the-way places. They need little attention as the dense heavy foliage is sufficient to destroy the vegetation attempting to grow below them.

S.wigstoniensis is a willow with a coppery-brown coloured bark. It prefers a rich loam soil and produces an osier known as Americana, from the fact that it has been extensively exported from Germany to America. It is long and supple and shows red eyes—a characteristic of the *purpurea* species. It has a moderate working quality.

Among other varieties which do not appear to be in use for osier production currently but which were so used in time past and may be revived in the future, are the following:

S.basfordiana Scaling ex Bean—the Basford willow. (It is believed that this willow first appeared in Britain about 1870

S. PENTANDRA The Bay leaf willow
(a) outline of form
(b) base of boles or stems

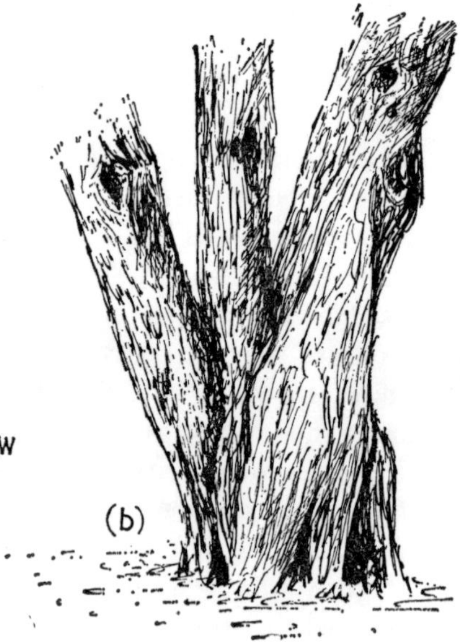

in the nursery of a William Scaling, who was for 10 years basket-maker to Her Majesty and the Royal Household of that period.)

S.decipiens Koch.	the cardinal or welch osier
S.forbyana Sm.	Forby's osier
S.hoffmaniana Sm.	Hoffman's osier
S.holosericea Willd.	the velvet osier
S.rubra Huds.	the red osier
S.smithiana Willd.	the silky-leaved osier
S.villarsiana Rouy.	Villar's osier
S.virescens Forbes	the verdant osier

While there has been a tendency in recent times for the osier business to decline, this has been due mainly to the universal increase in willow substitute materials, but it is reasonable to assume that, with improved economic conditions, the outlook for this age-old and fascinating business may yet improve.[7]

WILLOW SPRAYS FOR HOUSE DECORATION

Certain types of willow produce colourful and aromatic stems, foliage and catkins. During the winter and early spring when other decorative sprays are scarce and, consequently, expensive, such willows can be cut and despatched to the large city markets where they find a ready sale at good prices. The beauty of some species is not confined to the winter months but is present at other times too. This use of willow is exploited to the full in the United States of America and Scandinavia but there is room for an expansion of the trade in Britain.

The following varieties of willow are among others producing colourful sprays suitable for the commercial markets:

S.acutifolia Dahl.—violet-purple shoots with white bloom. Catkins in December.

S. PENTANDRA

Leaves and male catkin

S.argentea Sm. (variety of *repens.*)—silvery-grey foliage, yellow catkins in great profusion.

S.aurea (var. of *S.alba* L.)—pale yellow leaves and golden coloured branches.

S.balfourii Linton—broad grey leaves, long attractive catkins.

S.basfordiana Scaling—glossy orange-tipped twigs. Drooping catkins.

S.bockii Seemen—catkins crowded and conspicuous in late summer and autumn.

S.candida Willd.—leaves snow white and cottony.

S.chermesina britzensis Hartig.—young shoots brilliant orange-scarlet.

S.chrysostella Stokes—bark brilliant sealing-wax red in winter.

S.cinerea atrocinerea tricolor Dipp—leaves yellow on white or on white and red, twigs and buds are blackish.

S.cramacile Kukome—black catkins in late February.

S.crispa Forbes—catkins small, March and August; anthers red, later turning yellow.

S.daphnoides Villars—distinctly ornamental, yellow catkins from large crimson buds.

S.erdingerii Kern—purple-brown 'bloomy' shoots, attractive catkins.

S.glandulosea Setsuka—fascinating shoots, contorted and fasciated in unusual manner.

S.gracilistyla Miquel—yellow-grey and red catkins, shoots with woolly grey hairs.

S.pellita Anderss.—branchlets with heavy purple 'bloom'.

S.pentandra L.—densely flowered golden yellow catkins, late May or June.

S.pomeranica Willd.—preceding year's shoots covered with a violet-coloured 'bloom'.

S.proteafolia Schl.—young twigs, yellow-purple, covered with hairs.

S.purpurea L.—elegant, slender twigs, purplish hue on young shoots, catkins and anthers are red before they burst.

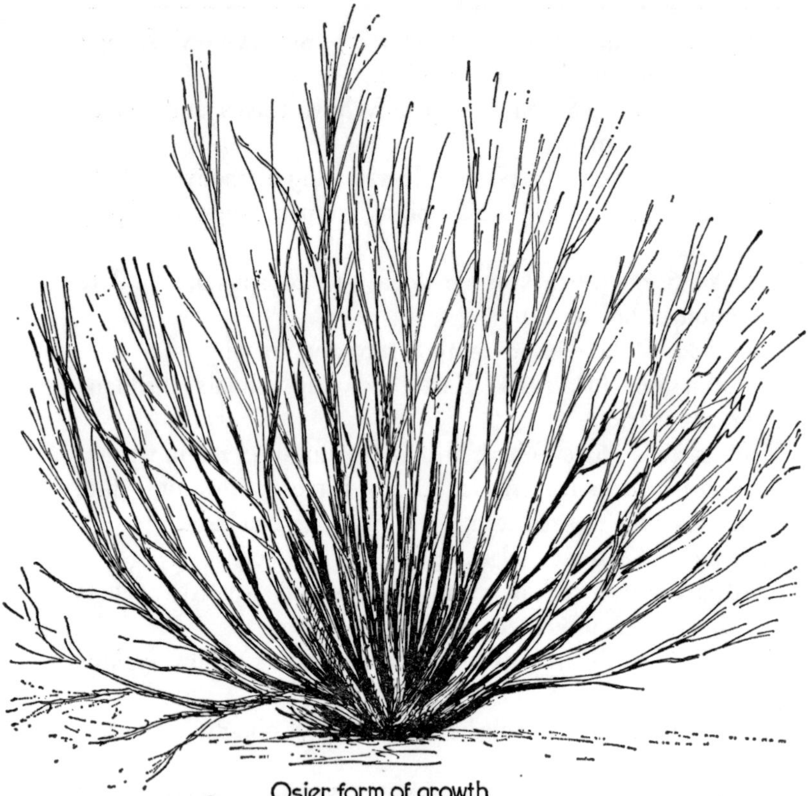

Osier form of growth

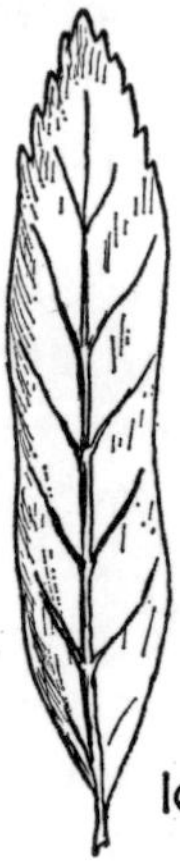

leaf

S. PURPUREA

The purple or osier willow

S.pyrifolia Anderss.—lustrous branches, shining red buds and red-brown shoots in winter months.

S.vitellina Arcangelli—thin twigs and branches bright yellow or orange.

S.wehrhahnii—masses of silver white catkins, giving the effect of icicles.

MISCELLANEOUS PRODUCTS

Most other products of willow are so varied in character that reference to them is more for interest value, than as a record of any commercial involvement they may possess. The early Britons and the Romans utilised the twigs and branches of willow to the full. It is recorded that:

> 'the bending willow into barks they twine,
> then line the work with spoils of slaughtered kine.'

The purple willow—*S.purpurea* L.—is not only ornamental but its shoots are long, tough and flexible and as such they may be formed into any shape. A fence of this material is considered to be superior to that of wire for the exclusion of hares, rabbits, etc. Moreover, the bark and leaves of this species are so extremely bitter that animals will not touch either.

A list of medicinal plants prepared in AD 64, which remained a valuable work of reference for over sixteen centuries, involved many species of willow from which were extracted a group of glucosides, pure saliciposide, salicylic acid and salicin—a bitter crystalline principle. A black dye, used also for tanning, is a product of several species of willow. In Scandinavia the inner bark of several varieties is kiln-dried and ground for mixing with oatmeal for human consumption in times of scarcity. In Tartary, willow bark is macerated and the fibre, when separated, is spun into threads from which cloth is woven. Many fascinating and fragile craft specimens are preserved in the museums at Kew Gardens, Surrey, and of particular interest are the hats and fabrics made of fine strands of willow, woven together with great skill and precision.

Beavers (they are intelligent creatures) prefer willow wood to all other kinds. Willow leaves and young shoots are wholesome and nourishing to cows, goats and horses. In fact, a number of shrubby forms of the genus are used in a big way as browse in the western cattle country of America. It is said that when horses are fed on this food between August and November the animals will travel 100km (62 miles +) a day, without becoming in the least fatigued. Moreover, animal dieticians speak highly of the food value of willow. The roots of willow, apart from their prime importance to the life of the tree, provide a purplish-red substance, which, when used with cochineal, is used for the dying of egg shells at Eastertide.

For centuries, farmers have been in the habit of pollarding the timber willows at a height of about 2–3m (8–9ft) to produce shoots which give repeated crops of small poles every 5 or 6 years. These are cut and provide material for many useful purposes, including gate-hurdles and other fencing needs on the farm. At the present time quantities of willow boarding are made up into large crates for the packing of machinery for export. As the veneer of willow will stand bending at right angles this fact, coupled with its clean white colour and light weight, makes it admirably suitable for the manufacture of various styles and sizes of chip baskets, punnets and the like, used in the marketing of soft fruit and saladings.

Polo, skittle, coconut shy and other balls, all of which demand lightness in weight and power of resistance to smashing blows, are among other products, while small pieces of the timber are utilised for the making of toys of many kinds, also broom and brush-heads and matchboxes. Children's cricket bats, rounder's bats and sticks are made from third-grade willow wood when reasonably sound; rejected clefts and branchwood is generally acceptable for pulping into fibreboard and even certain qualities of paper and cardboard. Garden baskets or trugs (which may be of any size from 8in to 4ft long) are made partly from willow and partly from sweet chestnut (*Castanea sativa*), the pieces of cleft willow being steamed to enable them to be bent more easily to the curves required for fitting within the rim of the chestnut. The blue-on-white pottery decoration—familiarly known as 'the willow pattern'—is attributed to Ma Yuan (*circa* 1190–1225 AD) but the weeping willow, which is native to China, appears to be a most popular subject for the artists' brush. Moreover, the fine texture of charcoal sticks when derived from charred willow wood is more than adequate to satisfy the artists' most fastidious needs.[12]

Willow products will never end, it would seem, but this list must. What a miscellany, to be sure! It must be concluded, not by shopping baskets, or larger ones for the fishing industry; not by butchers' or bakers' baskets, or baskets for the Post Office, or even for cats, or dogs, or pigeons; yes, willow provides all these and more; but let us conclude on a higher note by including the willow rods used for the framework, or 'shapes' of the interiors of the bearskins of Her Majesty's Brigade of Guards.

While many of the earlier products of willow are now a matter of history, having given way to more modern materials, a timber possessing hard-wearing qualities, which dents rather than splits or breaks under the roughest usage, very light in weight; takes a good polish without objectionable odour or taint and is free from resinous matter, can scarcely be devoid of some marketable value, even today.[6 and 7]

5: *Cultivation and Production*

CULTURAL GUIDANCE INSOFAR as it applies to specific willows is given in the appropriate chapters where the plants are listed. Much of the direction given below will apply generally to all willows in cultivation for any purpose whatsoever. An essential feature in the dealing with problems which arise is speedy attention to the appropriate measures recommended.

Site and Soil

The coerulean timber willow is inclined to be rather fastidious, but, like most other members of the genus it is not difficult to cultivate when one knows its requirements. It prefers a heavy alluvial soil with adequate drainage. A most useful guide to suitable soil can be gained from the type of wild vegetation growing in the vicinity of the proposed site. A vigorous growth of reeds, sedges, wild iris, figwort or great water dock, indicate marshy land, having a water table higher than is safe for the purpose we are now considering. Areas having a dense growth of meadowsweet are suspect, but may be improved by drainage with open ditches. Favourable conditions are generally indicated when a dense growth of nettles and ground ivy are present, coupled with smaller quantities of comfrey, ladysmock, goose grass, cow parsnip, angelica, with a sprinkling of meadow sweet and willow herb; not all of these need be present at the same time.[8]

When in cultivation, timber willows have a strong preference for normal seasonal temperatures, alkaline-inclined soil conditions, that is, not less than 6·5pH and a rainfall of about 35in a year. The bat willow particularly has a strong aversion to the salt-laden atmospheres of coastal districts and also to high humidity conditions. The latter tends to encourage the growth of lichens on the bark which in turn retards the growth of the tree. In good cultural conditions the tree will reach 20m (60–70ft) or more in height with a girth of 1·3m (50in) at breast height in 12 to 14 years. It then reaches maturity and is felled for its timber. The best cultural situations for bat willows are on the banks of rivers and streams which pass through a reasonably level terrain. Here they are planted in single line, but, if the soil is right, the water table not continuously within 12 to 15in of the surface and the land area adequate for the sets to be planted at not less than 40ft apart, they will flourish in block or plantation form. River banks where silt has been deposited by floods make particularly good sites; moist soils resulting from areas exposed, such as when a river changes its course or breaks its bank, also create suitable conditions. Spring ditches on upland sites are often capable of producing fair quality timber, but in such circumstances the trees may take a little longer to mature and, therefore, command a slightly reduced price on maturity. In making use of such upland sites it should never be assumed that because the land grows good corn, first quality timber will necessarily result.

Manures, Mulches and Irrigation

The need for occasional manuring of willow grounds is plain when it is realised that the cricket bat willow, being an exceptionally rapid grower, impoverishes soil as quickly as, or quicker than, many cereal or root crops. Moreover, the space between the trees is often exploited for subsidiary catch-crops thus making the need for renewed fertility all the greater. Among the most suitable manures and fertilisers are manure from the farmyard applied every three years; an annual top dressing of the remains of weeds dredged from rivers, and dressings from effluent ditches on sewage farms;

all have the effect of increasing the humus content of the soil. Occasional top dressings of wood-ash, chicken manure or stable dung, are all particularly beneficial. A tonic in the form of nitro-chalk, super-phosphate and potassium sulphate, in the proportions of 4–2–1, applied at the rate of 0·23kg (8oz) per tree or bush is strongly recommended. Where the land is inclined to become acid a little lime is helpful to return it to a 6·5pH reaction—bearing in mind that only a little lime is needed, any excess being harmful. It should always be remembered that the two chief factors in renewing and maintaining soil fertility for willows is an adequate supply of nitrogen and phosphorus.

Propagation and Planting

The most popular method of propagation is carried out by means of unrooted sets which are tall shoots up to 5m (15ft) in height, bearing their natural heads or crowns. The sets are cut from stools or 'tods', which are trimmed branches cut from mature trees and planted about 2ft deep and 3ft out of the ground. They are planted in set beds or areas specially set aside for the purpose.

Rooted sets are also on the markets; these are grown from short cuttings, produced in nurseries and dealt with in a manner akin to forestry practice. Sets for propagation are most successfully planted during December or, preferably, February or March. It is unwise to delay until April, unless copious watering is possible, and equally unwise to plant as early as November unless cultural and climatic conditions in the particular locality are found by experience to be normally favourable.

The aim of the bat willow grower is to produce fast-grown timber and this is achieved by an adequacy of moving water, nutrient salts in the soil for the root system and light and air for the crown. When grown in large numbers they should not be planted less than 11m (35ft) apart in line, increasing to 13m (44ft) when planted in blocks or plantation form. From the point of view of regularity of income and utilisation of space, it appears to be advantageous for the crop to be planted at such times as will result in trees being of uneven age, thus permitting a partial felling each year or alternate years, which would also make possible a slightly closer planting distance while still permitting a free crown development.

Before attempting to plant willows on the banks of water-courses, growers are advised to acquaint themselves with any bylaws of river boards which might be in force governing the spacing distance between the trees and also distance from the brink of the river or stream at which they might be placed. The normal method of planting unrooted sets is to remove with a spade the top spit of soil, then, with a crowbar, to drive a hole for a further depth of not less than 45·72cm (18in). The hole may, of course, be made with the aid of a mechanical boring tool. Into this hole the set is carefully but firmly embedded, and the topsoil replaced, care being taken to firm it round the base of the set. The firming should be continued at intervals, once or twice weekly during the first two or three weeks, and particularly after heavy winds.

It is desirable that this treatment should be continued during the growing seasons for the first two or three years at least. The planting of rooted sets differs from that of the unrooted in that it resembles more the method employed in normal forestry planting practice. In this case the roots should not be planted too deeply, the stub from which the set grew being just visible above the surface of the soil. It is desirable that the hole should be large enough to take the roots adequately spread out, every care being exercised to avoid damage to them. Like most other young things, bat willow plants need all the help they can be given, and at this stage, some protection by the firming of the soil around the base of the sets, is most valuable. Mulching with cut herbage, which prevents both drying out and freezing of the roots, also helps to form a good tilth, stops undue compaction of the surface, and, to a certain extent, acts as a fertiliser.

Bat willows are generally expected to provide timber of three bat lengths but where both the set and cultural conditions are approaching ideal, four bat lengths may be produced. The stem of the young tree should be as free as possible from knots, blemishes and damage of all kinds. One means of ensuring this is to start with sets which are clean in this respect and to continue to keep them clear of shoots right from the time of planting until the shoots cease to appear, which is round about the fourth or fifth year. The buds should be rubbed off by hand at frequent intervals, up to 2·4m (8ft) for 3–bat lengths, and to 3m (10ft) in the case of 4–bat lengths, the aim being to prevent the shoots becoming old and woody, and thus requiring the use of a knife or cutting hook to remove them. Where the leading shoot appears to be 'running away', topping may be necessary, but this should not be carried out after the tree has reached three or at most four years of age and then not lower than at a point where the leading shoot measures 2–5cm (1–2in) in diameter. If the operation is left until later it will result in the growth of the tree being retarded, with consequent damage to the timber.

Not all districts are suitable for the production of the coerulean willow and it may be that other varieties of willow may be favoured for the manufacture of certain commodities. There is a small group of timber-producing trees that are a little less fastidious in cultivation than *coerulea* but which can well take its place, for commodities other than cricket bats. These trees consist of a group of segregates, hybrids of *alba* × *fragilis* and classified as *viridis* Fries. They comprise *albescens* Anderss. and *fragilior* Host., according to which of the two parents they bear the strongest resemblance. The third and last member of the group is an apparently distinct segregate, but again a hybrid of *alba* × *fragilis* and known as *viridis* variety *elyensis* Burtt Davy. It is frequently planted in East Anglia and in parts of southern England. It is said to have been brought into cultivation from the Isle of Ely in Cambridgeshire.

Cultural Features for Ornamental Types

When willows are brought into cultivation, having the effect of creating differences in conditions from those of their wild habitats, a perfectly natural re-action is for them to appear slightly fastidious in their cultural needs, but they quickly settle down. After all, it may well be that climate, moisture, drainage and soil values are all changed in some measure, resulting in a slight shock being sustained by the plant. However, such plants usually recover in a good rich loamy soil, reasonably damp situations, well drained and with chalk totally absent. Swampy or boggy ground, peat moss and lime-stone soils, land with a high proportion of sand and gravel and waterlogged areas difficult to drain—all these should be avoided, except for one or two of the sallows which are not at all fussy. Stagnant water is, however, anathema to them all.

Where the natural habitat conditions differ very considerably from those of the cultivation areas, the latter can so completely change the willow characters as for them not to be recognisable for the same species; this is due to the process of adaptation of the plant to its changed circumstances. For example, narrow leaves become broad and shaggy, while woolly ones become smooth and shining; plants only 1 to 2ft high attain a height of 6 to 9ft. The wood of the trees, whether of trunks or branches, or of young shoots, is smaller, harder, tougher and more compact and durable than those of willows produced in rich moist soils. If, as may happen, it is preferred to plant on upland sites which are inclined to be dry, this can be done so long as the soil is deep and permeable.

Propagation by Cuttings

Cuttings are fully ripened shoots taken in the autumn or winter. Rooted cuttings are obtained by preparing one year old wood about 1ft long, cut straight across the lower end and sloping at the upper end. About 9in of the cutting should be inserted in the soil perpendicularly and the soil pressed firmly to it, especially at its lower extremity. The reason for it being cut directly across at the lower end is that it may form an equal callosity all round it and, consequently, throw out an equal number of roots on all sides. The cutting is placed in the soil upright so that the roots may be principally formed at its lowest extremity—that makes for a more handsome and symmetrical plant than when the roots are protruded partly from the end and partly from the side. The roots protrude most when the soil has been pressed firmly against the cutting. If the soil has not been so pressed, probably no roots will be produced and the cutting will rot. The upper extremity of the cutting is cut in a sloping form merely to throw off the rain.

When willows are to be planted directly in their permanent quarters, cuttings may be made of two year old wood, about 2ft long and cut in a sloping direction at both ends. The cutting should be inserted in the ground, in a slanting direction for at least 10in of its length and made firm with the foot. Cuttings of the smaller kinds of willows, and especially of those which are somewhat difficult to strike, should be planted in a sandy soil, in a shady situation and kept moist. Cuttings of the winter's wood should be taken in the autumn, the buds will swell during the winter and grow with vigour in the spring. A second firming of the cuttings in the spring is advisable; they may have been loosened by frost or excessive wet.[16]

Propagation by Seed

Some willows propagate very readily from seed and where reasonably large numbers of plants are required this is by far the most economical method. The seeds are small and black and enveloped in a tuft of cottony matter. Their ripening date varies slightly with the species, but the majority reach this stage in May or early June. The seeds should be sown wide-spread in moist soil shaded from the sun and lightly covered with loose litter. If a heavy shower of rain should happen soon afterwards, or, if not, they are watered well, they will spring up in about 3 weeks. Plants of between 3 to 4in will be produced before the end of the season, at which time they can either be planted in the position in which they are intended to stay or planted out for one or two years to gain root growth and then transplanted into permanent situations.

Pruning, Grafting, Budding

Hard pruning, that is practically to ground level, for coloured bark effect, should be carried out in March. Pollarding should be done in February but only if really necessary. Certain willows produce their best results when grafted or budded—the former being carried out during March while the latter is best deferred until July.

Sinks, Troughs and Pans

Dwarf willows are in their natural habitat when trailing over and around large boulders on the high mountain sides and sometimes in the deep valleys. When brought into cultivation the rockery, complete with water-retaining soil, is the obvious situation for them. The smallest varieties—up to 5 to 6in in height—are happiest when planted in sinks, troughs or pans, with medium sized gravel covering the shallow soil in which they are set. In these containers they can be placed with good effect either on the rockery or on a flat stone wall, where they will be greatly admired.

Lime Reducing Method

Willows are mostly lime haters; if the soil has too much it can be decreased by treatments with humus-rich dressings, such as, leaf moulds, peat, old mushroom bed composts, and the like. When wishing to loosen the soil around willow plants care should be taken not to damage the roots; they are shallow rooted and one is restricted, rather, to pricking the top soil carefully with a small fork. The same advice applies when handling and planting—the roots are so easily damaged. Preliminary digging of the ground should be done at least a month or six weeks before planting takes place to allow the soil time to settle.

Protection Methods

The dark abnormalities in the form of growths to be seen during the winter months on the bare branches of female willows are commonly known as Witches' Brooms but sometimes referred to as galls. The growths are found on a limited number of types of willows, usually they are on *fragilis*, *vitellina* and *alba*, and very occasionally on *coerulea*, *triandra* and *cinerea*. Some attribute the cause to a virus, others to mites, while others offer a fungus of the genus Taphrina, as the cause. It would, perhaps, be safer to conclude at this time that the cause is unknown and neither does the abnormality appear to be a harmful one.

Wounds of all kinds are liable to cause defects in timber and branches and thereby reduce the attractive quality and commercial value of willow trees. In the case of the exotics, osier willows and other cultivated forms, the injured plants become ready hosts to all kinds of pathological pests which tend to reduce the speed of growth, if nothing more serious.

Mechanical injuries are caused mainly by farm implements—often through human carelessness, and occasionally by animals. The bark of young willows is sometimes tender and sweet according to species and consequently protection from grazing animals, rodents, rabbits, hares and coypus is essential. Young willows intended for the production of commercial timber or decorative sprays, should not be planted so close to other bushes or trees as to allow the overhanging branches to thresh the tender willow stems and cause injuries to the bark or new shoots. If this warning is not heeded and wounds do occur they should receive early treatment to ensure rapid healing and prevention of decay. Torn bark should be removed with a sharp knife to avoid further tearing, exposed wood being treated with a good lead paint. Small wounds may be treated with grafting wax or bituminous paint—the aim being to cover the exposed wood with a smooth waterproof surface which will not crack.

It would seem almost superfluous to emphasise that any protection methods should be effective, but my experience has been that large amounts of money are spent on ineffective measures, making the whole process little short of fruitless. Painting the base of the young stems up to a height of 76cm (2½ft) with Stockholm tar every two years will keep off rodents, but an equally effective alternative is to fit sleeves of spiral plastic. This latter method has the advantage of lasting for an almost indefinite period. In about four or five years the bark will have become tough enough to be immune from attack, allowing the plastic sleeves to be removed and used again on other young trees.

Protection against attacks by horses and cattle calls for a different technique. If the trees are planted on the sides of arable fields no such protection will be necessary. When in grazing meadows one has the choice of either individual tree protection or the continuous wire fence, the latter being appropriate when the trees are on the perimeter of the field. The fence should be strong and not less than 1·2m (4ft), preferably 1·5m (5ft) away from the line of trees. This has one drawback in that it incurs the loss of an appreciable area of grazing land, revealing the advantage of the individual tree method. For this, a strong and effective structure is required. In many instances one finds thorns and brushwood tied round the stems of the trees. This is usually ineffective, the protective material itself often injuring the tender bark. Another form of ineffective protection is the erection of a sleeve of wire netting 1·8m (6ft) high and 1·5m (5ft) in circumference, with two short stakes inside the sleeve and a few strands of barbed wire round the outside. This method is quite useless within a very short time, as it gets pushed against the tree by animals, resulting in damage to the stem it is intended to protect, also leaving it wide open to the ravages of animals.

The protection should be high enough and strong enough to prevent the highest horse from reaching the stem of the tree and be encircled with sufficient strands of barbed wire to prevent smaller animals, eg calves, from getting their heads through the strands. Such form of protection will require for each structure the following materials:

4 timber uprights 2·3m (7½ft) long × 8·9cm (3½in) × 6·3m (2½in); 4 timber horizontals 1·8m (6ft) long × 8·9cm (3½in) × 3·8cm (1½in) for connecting the tops of the uprights. About 11m (13yd) of barbed wire will also be needed. The scheme is to construct a roughly square enclosure 1·8m (6ft) high, with uprights 45·7cm (1½ft) in the ground, the top of the square having equal sides of 1·8m (6ft) each, tapering to a smaller square with equal sides of 1·3m (4½ft) each, at the base of the structure. Five or six strands of barbed wire are then wound round the wooden frame from near ground level to top of frame and secured by galvanised staples. This results in a simple but strong and effective structure.

The timber can be of seasoned oak, chestnut or conifer thinnings, which are not difficult to obtain at very small cost on large estates.

If all the materials have to be purchased at full market prices, the cost in the first instance would be in the region of £5 to £6 per frame (12–15 dollars), but if the timber is effectively treated the protective structure should serve its purpose for a period extending over the life of more than one tree, thus making the cost of production reasonable. Apart from willows in cultivation this form of protection is applicable to large trees of other genera growing in parks, meadows or orchards.

The Elements—Frost, Sunscorch, Wind and Drought

Indigenous tree willows, while not regarded as particularly frost susceptible, often suffer serious damage when exceptionally late frosts occur in low-lying ground. The trees have a long growing season, their buds beginning to burst in March, well before the end of the frosts. Damage often occurs to leaves, flowers, and young shoots. Stems of young trees and sets still growing on stools can also be seriously affected. The resultant abnormally rough bark patches called 'frost-bark' are not particularly dangerous, but if new wood has started to form, a frost-ring will occur at the beginning of the annual ring and frost canker may also develop. Little can be done to mitigate these conditions, except to attend to the drainage of the soil, which is synonymous with good aeration.

Sunscorch occurs in the form of a long strip of bark on the south side of the tree being killed, usually rendering the tree unsuitable for bat timber. As both frost and sunscorch generally affect recently-planted sets, or sets in their early years of growth, it is more satisfactory to replace them by new sets rather than attempt to nurture them to a doubtful maturity. If, on the other hand, it is decided to retain them, every effort should be made to avoid, for the time being, sudden exposure of the young and tender stems to the full heat of the sun, as for example, by the too-early removal of any large protecting trees.

High winds are not often a serious factor, although they do occasionally cause young trees to experience what is called 'wind throw'—a loosening of their anchorage in the ground. Generally, this occurs only where the plant has a particularly shallow root system, such as is induced by impermeable sub-soil or an exceptionally high water table. Where these conditions obtain, it is advisable to rescue the plant by securing it to a stout stake by means of one of several forms of ties available from most nurseries; normally, of course, staking willow sets is not necessary—they are anchored in the course of planting. Dieback of rootlets brought about by an adverse state of soil, can be checked in large measure by effective drainage. In excessively dry summers, involving drought conditions, it is desirable that where normal water supplies to the trees are diminished to danger point, other sources of supplies should be utilised, even by manual labour, if no other means are available.

Fire and Chemical Spray Damage

The risk of loss of willows from fire is mainly but not entirely experienced in set- or stool-beds. During the summer months the undergrowth may become very dry and combustible and a fire risk may be created by the burning of hedge and grass trimmings on the perimeters of stool-beds. Equal risk may occur where the public have easy access to points near the willows. Some degree of protection may be afforded by cultivation or the keeping clear of herbage between the stools. This is, of course, simply the exercise of normal good husbandry, but so often neglected. Trees growing alongside ditches bordering highways are occasionally damaged by fire running through the herbage. More disastrous results are sometimes incurred when burning up the straw on the stubble after the harvesting of corn. The burning should take place only when the wind is favourable, that is, blowing away from the hedgerow trees.

The wind-drift of chemical sprays is often the cause of defoliation and other damage and sometimes the death of many valuable trees. Damage is likewise affected when the spraying machines are cleansed by the stream sides; the poisonous residue is emptied into the stream, the water becomes contaminated and is taken in by the trees' roots with deadly effect.

Upside-Down Growth

Orthodox gardeners may not be interested, but just horrified perhaps, at what they are about to read. Other folk may be interested to a point. However, a particularly curious feature about the willow is that its roots are more readily changeable into branches and the branches into roots, than are the trees of other genera. All that is necessary is to take up a plant, turn it up-side-down and bury the whole of the crown of branches in the soil, leaving all the roots above ground. The larger twisted roots become the principal branches and preserve their general form; the young shoots produced by these take the form and appearance common to the species in its natural state. I have seen this done with a number of young plants of *alba* and some of its related forms, and the purpose of the exercise was to create a curiosity, and I feel this was achieved. It is all rather fantastic, inclined to the grotesque and I do not recommend anyone to waste their time repeating the process as I am all against the unnatural in nature. But the upside-down willows did grow!

6: Botanical Features

PLANT STRUCTURE

THE STRUCTURE OF a plant, and that includes a tree for this purpose, consists of two main features. First, the root which serves the dual purpose of anchoring the unit firmly in the ground and absorbing moisture and essential minerals from the soil; second, the shoot, consisting of a stem from which develop the branches, leaves and flowers. The stem structure consists of minute cells and inside the fibrous cell-walls are found a substance called protoplasm and a nucleus which is the centre of the cell's activity—the controlling element of the whole unit. It is because of this very complete set-up that a new tree or unit can be developed from a small detached portion, for example, a cutting. The cells increase by division—the nucleus also always dividing.

Growth depends on roothold, air, water and mineral nutrients. While the roots collect from the soil water and essential mineral salts which pass upward to the leaves through the inner layers of the stem, the leaves in turn collect carbon dioxide and water from the air, and, with the aid of a green substance called chlorophyll, evolved by the tree itself, these various substances are turned into a sugary sap, a carbohydrate, which passes downward through the tree just underneath the bark. The sap serves to develop and make new wood—hence the most recent layers are called sap-wood and are light in colour. The older or heart-wood, is darker in colour and consists of dead cells. The bark, also consisting of dead cells, is the hard inelastic protective covering for the delicate new or sap wood. In brief, therefore, the first essentials for a tree's development are a firm foothold in the earth, good supply of sunlight, plentiful supply of fresh air, adequate water available to the root system combined with appropriate soluble chemical nutrients in the soil. The nutrients, in the form of mineral salts, include nitrogen, phosphorus, and potash, with lesser quantities of calcium, iron, magnesium, chlorine, sodium and sulphur. Other vital elements are carbon, hydrogen and oxygen, all obtained from air and water. The available amounts of these mineral nutrients vary as between the differing soils, while the requirements as between tree and tree vary also, according to size and species.

Apart from adequate supplies of essential nutrients, the fundamental of a tree's life is its breathing, which means oxidation of carbohydrates in the tree's cell-sap so that they combine with oxygen from the air to form carbon dioxide gas. This process releases energy required for activity in its cells. In the process of assimilation of carbon in the presence of light, called photosynthesis, the earliest easily detectable products in green parts are sugar, starch or oil. During the first year each new twig or shoot increases in girth and length without making any additional annual rings. In succeeding years it increases in both length and girth and annual rings are produced. The reproductive organs are found in the flowers or catkins, the male organs or stamens carrying the pollen requiring to be transferred to the female carpels for the purpose of fertilisation.

IDENTIFICATION

The earliest record of plant identification occurred, *circa* 285–205 BC, when certain herbs, shrubs, trees, fruits and cereals were noted, together with their general habits of growth and uses to man. No system of botanical classification or naming of plants, as we understand it today, was evolved in those far off times. It was not until 1753 that the binomial system of nomenclature was established and this, with certain amendments, is in force today. By this time a great deal of knowledge had been gained including the distinguishing of willows by their magnitude, the shape of their leaves and the forms of their flowers. The knowledge that plants had sex and that pollen was necessary for fertilisation and the formation of seeds, was also discovered round about that same period.

The determination of the many sorts of willow is greatly complicated by the freedom with which the plants hybridise, by the sexes occuring generally, but not entirely, on different trees and also through the flowers appearing on some species before the leaves and with or slightly after the leaves on other species. The leaf shapes of the many kinds of willows are also varied, ranging from long and narrow, slender, pointed and toothed, to almost circular, with countless intermediate gradations. Occasionally they are found to vary on individual plants of the same species. It has been said that:

> Mother Nature is largely responsible for the many discrepancies we encounter as she has not differentiated her species clearly and that in no large class of plants do the types merge into each other and blur the dividing line so completely as they do in the willow.[1]

Moreover, the willows which inhabit low moist situations in valleys, flower only in the spring, while those of the high mountains do not flower until after the melting of the snows, which seldom happens before early summer. On the other hand, very many sorts in intermediate localities are intermediate also in their time of flowering. Hence, plants of the same kind of willow when inhabiting three different regions, as frequently happens, can offer three different stages of growth at a given time. This situation has resulted several times in three species being made out of one erroneously. Just another of the complications of willow! Willow also shows a great deal of variation from specific norms. This is due mainly to hybridisation between a limited number of morphologically well distinguished species. We are told that:

> it is not entirely clear why, with hybridisation frequent, the species as a whole has kept so distinct. Quite possibly

> they are of old standing and have been well sifted by natural selection to suit special environments. Owing to the activities of man these conditions are less distinct than they were when the plants were growing in more natural conditions; the creation of numerous but scattered environments of intermediate or synthetic types tending to bring species into juxtaposition, that crossing between them now occurs much more readily than it did in the past.[17]

Incidentally, some striking results have been obtained in Sweden in the building up of hybrid willows under fully controlled pollinations. The genes of one-third of the twenty-four species of willow occurring in Sweden were united into one plant and named *S.polygena*. The female shrubs were extremely fertile with the pollen from the male specimens, over 200 offspring of the original *polygena* having been raised.

In general, the responsibility for excessive hybridisation is placed upon the shoulders of nature, but there are some persons who take another line and shift some of the blame on to earlier botanists. Specific distinctions we are told have been made to such an extravagant pitch that no two persons were, or could be, agreed on what constituted a species, or what not, in *Salix*. There are some who are disinclined to give a distribution of the species of this genera, not because willows are in themselves uninteresting, for their variety of character and stature renders them otherwise—not because they are unimportant, for their numbers and wide distribution render them a prominent feature in botany—but because they have been rendered botanically odious in books.

When we study a number of the modern flora and note the confusions and contradictions which still appear to exist about willow we are at first inclined to take a rather similar view. However, we must not overlook the untiring labours over the years of so many great botanists in their efforts to resolve these and countless other problems, even though they have not been entirely successful all the time. After all, the tasks are not simple and the areas over which the problems arise are immense.

The following short synopsis of morphological characters of the willow will be found helpful when attempting identification:

Bark is rough to comparatively smooth, but should be checked as to whether persistent or flaking.

Leaves are simple, undivided, generally stalked, alternate and deciduous.

Each lateral resting-bud is encased in one bud-scale.

Many terminal buds on branches die in late autumn and drop off in winter. In following year the highest axillary bud on these branches shoots out and continues the branch growth. Thus willow branches are sympodial.

Leaf appendage (stipule) features are size, shape, marginal toothing and relative duration (persistent or deciduous).

Leaf stalk (petiole), its relative length, presence or absence of glands near apex.

Leaf blade, its shape and relative size, shape of apex and base, absence or presence and character of hairy covering (indumentum); size and relative number of stomatal dots on leaf surfaces, character of leaf margin—whether toothed or free (entire) and, if toothed, whether serrate or dentate and whether glandular or not.

Stamens and carpels do not usually occur on the same individual plant.

Flowers are numerous and carried in catkins which are subtended by scale-like bracteoles.

Catkins arise in July or August in the axil of a leaf on a part of the stem formed during that year. The leaves fall in autumn, consequently when the catkins burst out in the following year (March to June) they are seen in axils of fallen leaves on a part of stem produced earlier. In some species the inflorences open before the leaves emerge from their buds. The axis of the inflorescence bears at its base a few scales and higher up a number of scale-like bracts and axillary flowers. One flower stands in the axil of each bract. Relative length of catkins differ with species. Features to study are: thickness and density of flowering; relative length of peduncle; shape, length, colour and hairiness of bracteoles.

The staminate or male flower consists of two stamens and a greenish nectary situated on base of bract. Stamens have long filaments, extrose anthers and sticky pollen. The carpellary of the female flower is also inserted on a bract and consists of a nectary and a syncarpous gynoecium composed of two carpels. The ovary is stalked in some species but stalkless in others. It has one chamber containing many ovules attached to two parietal placentae. In some species stigmas are borne on a stalk, in others the stalk is absent or nearly so. The nectary is regarded as part of the flower and, being inserted below the ovary, the flower is described as hypogynous. The number and shape of glands at base of ovary is useful for identification of a species. The fruit is a two-valved capsule which allows the escape of numerous minute seeds. The seeds are scattered mainly by the wind, and each seed is possessed of a tuft of silky hairs which forms the sailing mechanism.

The following guide may prove helpful when attempting to identify incomplete specimens. (Leaf-characters refer to *adult* leaves unless otherwise stated. Catkin-characters refer to the *female* catkins unless otherwise stated):

Plant procumbent, with stem appressed to the ground:

S.*reticulata* L.
S.*retusa* L. group.
S.*rotundifolia* Trautr.
S.*myrsinites* L. group in part.
S.*arctica* Pallas.
S.*pyrenaica* Gouan.
S.*reptans* Rupr.

Stems underground: (i) mountains and arctic:

S.herbacea L. | *S.polaris* Wahl.

(ii) bogs, swamps or on sand:

S.myrtilloides L. | *S.repens* L. group.

Branches with a bluish, waxy bloom:

S.daphnoides Vill. | *S.acutifolia* Dahl.

Decorticated twigs with prominent ridges:

S.bicolor Willd.
S.aegyptiaca Boiss.
S.cinerea L.
S.aurita L.
S.atrocinerea Brot.

Leaves narrowly linear and more or less entire:

S.rosmarinifolia Wimm. and Grab.
S.caspica Pallas.
S.wilhelmsiana Bieb.
S.viminalis L. group.
S.elaeagnos Scop.

Leaves very glossy above:

S.pentandra L. | *S.glabra* L.

Leaves regularly blackening on drying:

S.nigricans Smith
S.glabra Scop.
S.purpurea L.

Leaves glabrous even when young:

S.pentandra L.
S.purpurea L.
S.glabra Scop.
S.triandra L.
S.fragilis L.
S.caesia Vill.

Leaves with appressed silk indumentum:

S.alba L.
S.glauca L. group.
S.cantabrica Rech.
S.repens L. group.
S.viminalis L. group and hybrids.

Leaves with rust-coloured hairs beneath:
S.atrocinerea Brot., and hybrids.

Stamens free: most species, except:
S.purpurea L. *S.caesia* Vill.
S.amplexicaulis A & EG Camus.

Stamens united:
S.purpurea L. *S.caesia* Vill.
S.amplexicaulis A & EG Camus.

Stamens 2: most species except:
S.pentandra L. *S.triandra* L.

Stamens 3:
S.triandra L.

Stamens more than 3 (usually 5):
S.pentandra L.

Ovary glabrous:
S.herbacea L.
S.retusa L. group.
S.rotundifolia Trautz.
S.lanata L. group.
S.phylicifolia L.
S.nigracans Sm. group.
S.glabra Scop.
S.crataegifolia Bertol.
S.silesiaca Willd.
S.repens L. group in part.
S.hastata L.
S.pyrifolia Anderss.
S.elaeagnos Scop.
S.daphnoides Vill.
S.acutifolia Dahl.

CLASSIFICATION

Because willow plants bear seeds containing an embryo they are placed in the botanical Division SPERMATOPHYTA and as the seed is covered or concealed in an ovary and not exposed it becomes a member of the sub-Division ANGIOSPERMAE.

Incidentally, the Angiosperms are the commonest, most complex and most widely distributed plants inhabiting the surface of the earth. By reason of the willow having netted-leaf venation and fibrovascular bundles in the form of a ring or rings round a vascular pith, it is now allocated to the Class DICOTYLEDONEAE. As both sexes of *Salix* have aments or catkins the genus is placed in the sub-class AMENTIFERAE from which it passes into the Order SALICALES.

Up to this point *Salix* has been one of many genera which together have made up each of the several foregoing categories. It now joins with two other genera to form the Family SALICACEAE. (The two other genera are the poplar—*populus* L., and a plant consisting of a single species named *Chosenia* Nakai, after Chosen, the Japanese name for Korea, but for all practical purposes *Chosenia* is now united with *Salix*. With the poplar we are not here concerned.) From this point onwards *Salix* assumes its complete independence and becomes the sole member of the Genus SALIX. The Classification from this point, assumes a different purpose, namely, that of breaking down the *Salix* genus into several constituent parts, viz: species, sub-species, varieties, hybrids and forms. Any one of these parts may be divided still further, but it is scarcely necessary here to enlarge upon this extension of divisions.

The generally accepted unit in botanical classification is the *Species*, which has been described as 'the perennial succession of similar individuals perpetuated by generation'. The foregoing is the generally accepted international botanical classification, so far as *Salix* is concerned, but there is another method of grouping plants based on their mode of origin or their status in a particular country. It was introduced in 1847 and the choice of its constituent plants for each of the four categories is believed to be substantially accurate today.[19] The first group comprises plants whose presence in a particular country is in no way due to human action, either direct or indirect. The natural vegetation of a country is made up almost entirely of this category and these are sometimes referred to as 'the aboriginal possessors of the soil'. Members of this group are known as NATIVES.

The second group called COLONISTS consists of species owing their presence indirectly to the activities of man. In the absence of agriculture they would find no home for they are nothing more than the weeds of cultivation.

Next come the plants which have been deliberately introduced into the particular country by man for the purpose of cultivation, but which have subsequently escaped and established themselves amongst the natural vegetation of the wild and become a permanent feature of it. These are designated as DENIZENS.

Lastly come those which are introduced by accident through the medium of seeds or fruits. These flourish for one or two seasons, but seldom reproduce. They appear only in disturbed ground, play no part in the natural vegetation and are therefore transitory. Owing to their rather foreign origin, they were called ALIENS, but more recently the terms CASUAL or ADVENTIVE, have come into use, according to choice.[9]

NOMENCLATURE

The true derivation of plant names is often doubtful and the suggestions here offered for *Salix*, while being as accurate as possible are probably not entirely free from error.

The term *Sallis*, is reputed to be Celtic and in its analysis is *Sal* = 'near' and *lis* = 'water', in reference to the willow's general habitat. The term *Salire* comes from the verb 'to leap' on account of the extraordinary rapidity of the plant's growth. As the willow is practically a world-wide genus, a list of a few synonyms by which it is sometimes known in other languages may be of interest. While the names, willow, willowy, withy and sallow are believed to be among the oldest in the English language and with derivations uncertain, there is little doubt that they have their counterparts in the following:

Anglo-Saxon	Withig	Italian	Salcio
Dutch	Wilg	Latin	Salix
Flemish	Welge	Norwegian	Selje
French	Saule	Scottish	Saugh
Gaelic	Seileich	Spanish	Sauze
German	Weide and Felber	Swedish	Wide
		Welsh	Helyg
Greek	Itea	Hebrew	Harab

The common or colloquial names of the individual members of the willow genus are not regulated by any constituted authority and, in the main, are derived from a variety of sources, eg locality or region, commemoration of persons, adaptation of names, distinctive features of the particular plant, its habitat, and so on. It will readily be seen that the use of such names in an uncontrolled way could quickly result in a state of confusion in plant nomenclature, as, indeed, it once did. The antidote to this confusion was found in the binomial system introduced in 1753.[10] Not only did this system pave the way for scientific names to pass into common use but the system gave each plant a 'generic' or group name followed by a second or 'trivial' name. The trivial was quickly changed to a 'specific' designation which, when added to the generic comprised the binomial name. In practice the binomial should be followed by either the full or abbreviated name of the person responsible for the published description of the plant, eg *Salix alba,* Linn. or L. (for Linnaeus) (The White Willow); or *Salix adenophylla,* Hook. (for W. J. Hooker) (The Furry Willow).

Today the International Code of Botanical Nomenclature ('Botanical Code') governs the use of botanical names in Latin form for both cultivated and wild plants. The aim of the Code is to promote uniformity, accuracy and fixity and to ensure that a precise, stable and internationally acceptable system is available for such names. Thus we have under international control and management an almost effective system of naming the world's plants and at the same time, fortunately, we are still able to append common or popular names, with all their charm and appropriateness, to any plant we please. For example, we can still refer to the 'pussy willow' when of course, we mean the goat willow or sallow or, more scientifically, *Salix caprea* Linn. Or, again, in conversation with the 'locals' concerning, say, *Salix fragilis* L., we happily refer to 'the old crack willow' and are clearly understood. We know that whichever name we use, scientific or popular, according to the occasion or the company, both are correct—so long, of course, as we both are, in fact, speaking of the same willow!

7: *Insects and Other Pests*

All the attractive and ornamental features and commercial values which I have outlined are set at nought if the many insect pests to which willows are prone are not controlled or eradicated sufficiently early to be effective. This does not imply that willows are more susceptible to attack than are other plants, it simply means that they have their fair share of the problem. Unfortunately, methods of control do not at present cover the whole gamut of the attacking force but as many as are known are here given. The pests fall into fairly well defined groups, namely, (i) those which feed on, or cause galls upon, the leaves, main stems, shoots or branches, and (ii) those which burrow into the wood itself.

I will now give particulars of some characteristic injuries, together with details of the most likely pests responsible. Appropriate methods of control, so far as are known, are given at the end of this information. The four headings under which the information will appear are: aphides; moths; beetles and weevils; midges, sawflies and mites.

Aphides

Rods and stems covered with black fly or blight, the insects closely crowded together and sometimes killing the rod.	Willow Aphides (especially *Melanoxantherium Salicis*)
Leaves and young shoots covered with black or green fly or 'blight', or with sticky honey dew and black mould.	Willow Aphides (various species)

The willow is not immune from attack by colonies of aphides or plant lice, the more usual species being *Cinara saligna*. These are large dark brown aphides of both winged and wingless forms. A feature of an attack is the large amount of honey-dew produced. This falls on the foliage below, coating the leaves with a kind of varnish and attracting large numbers of wasps which feed greedily upon it. The honey dew, a product of the aphides, accentuates the damage done by the insects themselves. Often the damage to willows by aphides is simply discoloration and premature leaf-fall, while at other times trees as large as 30 to 40ft in height may be killed. The wood is often stained a dark brown colour where the insects have been feeding and, in the case of basket willows, the rods become useless for their intended purpose. Certain of the more common species, eg *Siphocoryne capreae* and *S.pastinaceae*, live partly on the willow and partly on members of the *Umbelliferae*, such as hemlock, wild parsnip, chervil, angelica, and for obvious reasons efforts should be made to eradicate such weeds. Species of aphides common on the leaves and shoots of osiers are *Aphis saliceta*, the two species above-mentioned, and *Pterocoma pilosa*. The most common species on the rod itself is *Melanoxantherium salicis*. The weeping willows are more prone to attack by aphides than are any of the other ornamental types.

Moths

Terminal or end bud of shoot tied up with a few fine strands of silk and often containing a brown chrysalis or a small caterpillar that feeds on the young growing leaves.	Small Willow Moths (various species)
Stems, mainly when of two or more years' growth, with	Clearwing Moths

a channel bored up the centre near the base, sometimes containing a white grub or beetle.	
Boring under the bark then within the wood making a vertical tunnel 6–8in long. Chiefly affects the crown and branches rather than the main stem.	The Leopard Moth (*Zeuzera pyrina*)
Leaf defoliators of timber willows.	The Puss Moth (*Dicranura vinula*) The Buff-tip Moth (*Phalera bucephala*)
Tunnelling in the bark and sapwood at the tree base and in the roots.	The Hornet Clearwing Moth (*Trochilium bembeciforme*)
Boring and tunnelling through the bark into the wood of the basal part of the trunk of timber willows.	The Goat Moth (*Cossus cossus*)

Damage to willow by the Goat Moth, *Cossus cossus*, is often a transitory nature, but when severe it is important as it involves the basal portion of the trunk. The Goat Moth is one of Britain's largest, the female has a wing span of 9cm (3½in). The caterpillar, full grown, measures over 7·5cm (3in). It has a dark brown head, the dorsal surface of the body being purplish brown and the rest of the body flesh-coloured. In the early stages it bores between the bark and the wood, after which it enters the wood. The tunnel in the wood is irregular, and the duration of the larval life varies from 2 to 4 years.

Another of our large moths, the leopard, *Zeuzera pyrina*, also attacks willows. It has a wing span of between 5 and 8cm

(2–3in) and is easily recognisable by its white wings marked with bluish black spots. The caterpillar measuring from 4–5cm (1½–2in) when fully grown, is yellowish white, with three dark spots, one on the dorsal surface of the head, another on the first thoracic segment and a further one on the last abdominal segment. The eggs are laid on the stem and branches of the willow from June to August. The caterpillar bores first under the bark, then within the wood, making a vertical tunnel 15–20cm (6–8in) long. The life-cycle varies from 2 to 3 years, and the damage chiefly affects the crown and branches rather than the main stem. The eggs of the puss moth, *Dicranura vinula,* are laid in May and June, caterpillars appearing a few days later. They feed on the leaves until September and when fully grown are 5cm (2in) long and easy to see, therefore hand picking is an easy method of control. The larvae of the buff-tip moth *Phalera bucephala* can defoliate both old and young trees of the cricket bat variety, but the young trees suffer most. The caterpillars are yellowish with black longitudinal lines and are covered in hair. The moths have a 6–7cm (2½in) wing span.

Further investigation is required before the life-history of each species can be described in detail. The larvae first become noticeable in the late spring when they spin together a few leaves at the tips of the growing osier rods, forming small tubes or nets, one larva only being found in each shoot. They feed on the growing buds and leaves, and when full-fed change into brown chrysalides in the nests, from which in due course the moths emerge. The damage prevents the rod from attaining its full length and also, owing to the destruction of the growing point, encourages the production of lateral shoots, a condition known as 'bushy top'. The most common species in the Somerset district are *Hypermecia cruciana* and *Depressaria conterminella.*

There are two species of the Willow Clearwing Moth of which the first, *Trochilium bembeciforme,* the Willow Hornet

Clearwing, so closely resembles a large wasp or hornet that it may readily be passed over. The second species, *Sesia formicaeformis*, the Red-tipped Clearwing, also has little resemblance to a moth, and is perhaps more like an ichneumon fly. The larvae of these moths are white grub-like caterpillars with brown heads and eight pairs of legs, which are small but evident. The Hornet Clearwing larva lives inside the stumps and especially in the rods of two years' growth or more. The larva of the Red-tipped Clearwing lives mainly in the stumps. The larvae of both species pupate in the spring in the burrows, and the moths emerge in June and July. The damage done by these insects is not very evident unless a few stumps are cut open, when it is often found that the wood is tunnelled in all directions by the larvae, which at first reduce the productiveness of the stump and later kill it altogether. These two species are chiefly responsible for the decay of stumps in osier beds.

The species of clearwing moths which attack tree willows is the hornet clearwing, *Sesia (Trochilium) apiformis* Cl., which has a wing span of 6·35cm (2½in) and a yellow and black scheme of coloration, including a black and yellow banding of the abdomen. The caterpillars tunnel in the bark and sapwood at the tree base and in the roots, the life-cycle lasting two years. Destruction by cutting and burning of infested stems is usually the most effective control measure for all wood borers. As the hornet clearwing burrows in the roots, the latter also must be destroyed.

Beetles and Weevils

Young rod in early summer broken as if by wind, but close examination shows a hole or puncture at the point where the break occurs.	Willow Weevil (*Cryptorrhynchus lapathi*)

Leaves and shoots eaten by dark green or blue beetles (which fall but do not jump when disturbed) or by their larvae, which are blackish or yellowish grubs, sometimes known as Army worms.	Common Willow Beetles (*Phyllodecta vitellinae* and *Phyllodecta vulgatissima*)
Similar damage, but light brown beetles present.	Galerucella Beetles (*Galerucella lineola*)
Stumps and stools weak and dying; when split open disclosing burrows and channels, containing white grubs, chrysalides, or beetles.	Musk Beetle (*Aromia moschata*)

In the adult state, willow beetles are shining dark green or blue. They first appear in the spring and early summer and at once attack the developing shoots and leaves, causing great injury by eating into the growing point of the rod. Eggs are laid in groups on the under side of the leaves and in a short time produce small, dirty grey or yellow coloured grubs or larvae, which at first remain together eating away the underside of the leaf. Later, as they grow larger, they spread to the other leaves, devouring each so that only the upper paper-like cuticle is left, and in severe attacks, all the leaves on the plant are killed. When full fed, the larvae fall to the earth and change into pupae from which the beetles are produced. There are two broods of beetles in the season, but the generations overlap somewhat, so that the beetles and larvae are often found together.

The beetles of the second brood leave the rods in autumn and crawl into dry places, such as heaps of rubbish, under the bark of, and into the crevices in, pollard willows, and shelter there in clusters during the winter, reappearing to attack the willows again the following spring. Control is difficult as natural enemies are unknown. Among the tree willows *S.fragilis* is sometimes attacked, but the cricket bat willow not at all.

Beetles of the genus *Phyllodecta* are common all over the country, and all willow-growing areas are subject to their attacks. *P.vulgatissima* is specially harmful to *Salix viminalis* in the Midlands and East Anglia. It does not appear to harm varieties of *S.triandra*. *P.vitellinae* attacks varieties of *S.purpurea*, also *S.alba vitellina*, but does not damage *S.triandra*. The Galerucella Beetle (*Galerucella lineola*) in the adult state is a yellow-brown beetle with dark markings on the upper surface. In its life-history and habits, with the exception of minor differences, it resembles the *Phyllodecta* beetles just described. It appears, however, to spend the winter in damper places and is especially harmful in the Somerset district where it attacks varieties of *Salix triandra*.

The willow weevil (*Cryptorrhynchus lapathi*) on account of its long trunk or proboscis, is sometimes known as the elephant beetle. It is partly blackish-brown in colour and partly yellow (or pink when freshly emerged). It has a life-cycle of two years. The adult beetle appears from early summer onwards, and may be found clinging to the stem with its trunk buried in the soft growing portion, which subsequently bends over as if broken by the wind. At the smallest disturbance the beetle falls to the ground and remains motionless, looking very like a bird-dropping. Eggs are laid in the stump or rods and produce white, grub-like larvae, which burrow in the stumps and sometimes up into the rods of two-years' growth. *Cryptorrhynchus* larvae have no legs, which distinguishes them from the larvae of the Clear-wing Moths, and they are round and sometimes short and so differ from the larvae of the Musk Beetle, which are rather long and flat. When full-fed they pupate in the burrows; the beetles emerge from the pupae in the autumn, but seem to remain in the burrow until the following spring. They are partially responsible for the decay of stumps in osier beds.

This beetle attacks alder as well as willow, and is common in all willow-growing areas, particularly Somerset and East Anglia. It is a serious pest, for, in addition to the annual destruction of a large number of rods and sets by the adult, great injury is done to the stumps and stools by the larvae. It is strongly suspected of being an agent for the spread of the vascular bacterial disease *Erwina salicis* (Watermark Disease). The Musk Beetle (*Aromia moschata*) is a large, shining blue or green beetle with long antennae. When disturbed it gives out a strong musky smell. The larvae is a large, white grub, legless, and rather long and flat in shape. It feeds in pollard willow trees and in old willow stumps, and incurs considerable damage through its boring. Adults can be seen in flight during the month of September. It can hardly be regarded as a serious pest when willows are well grown, but it is worthy of mention as it is so large and conspicuous that it is apt to excite interest. Apart from feeding on the willow *Curculis (Balanobius) salicivorus,* like *Cryptorrhynchus lapathi,* has been shown experimentally to be capable of transmitting *Erwinia salicis* (Watermark Disease) into healthy willow, presumably by the punctures it makes on the terminal part of young shoots.

The larvae of the woodwasp *Xiphydria prolongata* (dromedarius) bore into the timber, particularly that of dying or recently dead cricket bat willows. Its life-cycle is annual. One of the pests responsible for wood boring in the willow is the small poplar longhorn beetle, *Saperda populnea,* but while this beetle causes galls on poplars it seldom does so on the willow. It does, however, weaken the young stems at the point of attack where breakage often occurs. The larvae tunnel under the bark, quickly entering the wood and finally penetrating to the pith. The larvae are much sought after by woodpeckers and are also attacked by insect parasites.

Midges, Sawflies and Mites

Leaves eaten by livid blue and orange caterpillars.	Willow Sawfly (*Nematus salicis*)
Leaves with yellow or red lumps on them.	Willow Gall Sawflies (*Pontania gallicola*, etc)
Rods or stems, especially of two years' growth, with minute pinholes through the bark near the base. Inside are small burrows often containing minute orange-red grubs.	Willow Wood Midge (*Cecidormyia saliciperda*)
Terminal or end bud in the form of small rosette or button, often known as 'button top.	Willow Gall Midges (*Cecidormyia rosaria* and *Cecidormyia heterobia*)

Rhabdophaga terminalis, the bat willow gall midge, is a small fly belonging to the diperous family Cecidomyidae, and is particularly damaging to willows. The larvae attack terminal leaves, curling and wrinkling them, thus preventing proper unfolding. The gall is reddish at first, turning black later. Blister galls also occur on the midribs. Attacked terminals cease to grow. Where the attacks are on bat willow sets, side shoots form on the young rods at heights between 2 and 3m (7–10ft) constituting a serious form of damage as the intention is to keep the stems as clear as possible from side growth.

The willow wood midge, *Helicomyia saliciperda* Dufour, is a tiny fly which in the spring lays its orange-yellow eggs, chain-wise, on the bark of willow branches. The larvae, which are yellow or orange-red, bore into the bast region, and overwinter there. The damage is obscured at first by the thin bark, but this soon breaks up, revealing the punctured sapwood. Suc-

Page 137 (left) *Cricket-bat or the coerulean willow, in rural site in East Anglia*

(right) *the master-craftsman; his knowledge of willow timber far exceeded that of all others; he made more cricket bats by hand than any other man; his name was known wherever cricket was played or bat willows were cultivated. He was the late Walter Warsop, and is here shown in his workshop at Danbury, Essex, England*

Page 138 (above) Growth on stools at the end of the third year, with sets being cut and bundled preparatory to delivery to planting sites; (below) growth on stools at the end of the first year

cessive attacks gradually extend over the injured area, the stem often being encircled by the scar. Many willow species are attacked, those growing on poor soil, and thereby having less powers of resistance, receiving the greatest attention. Application of a lime wash will discourage egg-laying, but destruction of the trees by cutting and burning is by far the best control measure.

C.heterobia lays its eggs on or near the growing point of the plant, and the larvae or grubs burrow into the buds, as many as thirty occurring in one bud. The attacked buds fail to develop normally and become bunched together or form a distinct rosette, giving rise to the condition known as 'button top'. The terminal buds are those most favoured by the midge, but lateral buds and male catkins may also be attacked. There are two main broods in the year but these overlap considerably, so that adult midges are on the wing almost continually from April to September. When full-grown, the grubs normally turn to pupae in the 'buttons', but if the rods are cut in the autumn or in winter the grubs leave the galls and pupate in the soil. In winter, too, large numbers are scattered by tits, which destroy the galls in search of the grubs. The grubs that escape the attacks of the birds also pupate in the soil. The injury done by these insects is quite serious, since attacked rods are stunted and may also be 'bushy topped'.

There are several other species of gall midge that attack willows, but mention need only be made of the Willow Wood Midge, *C.saliciperda*. The grubs of this insect live in burrows in the stubs or in the sticks near the base. They pupate in the burrows, and the flies emerge through minute pin holes in the bark. The winter is spent in the larval condition in the wood. Attacked sticks, which are usually of two years' growth, are of little value as the attached portion is weak and must be cut away. The effects of attacks by both *Rhabdophaga resarea*—the European rosette willow gall midge—and the midge-like fly *Cecidormyia rosaria*, are to cause cessation of terminal growth and formation of lateral shoots.

Croesus septentrionalis and *Nematus salicis* are the two sawflies most commonly found on willows. When feeding, the caterpillars have the habit of curling the body either over their heads or round the plant below them. They are conspicuous on account of their bright coloration. They feed from July to October, chiefly attacking *S.alba, S.fragilis* and *S.caprea*. The caterpillars are only slightly attacked by birds on account of a distasteful fluid which they eject as a form of defence. Parasitic insects sometime effect a certain amount of control. Hand picking of eggs or larvae can be practised or the caterpillars, when in the young state, can be arsenically sprayed. N.*salicis* causes much loss on the Continent and occasionally in Britain. The larva is a livid, blue and orange caterpillar. It feeds on the willow leaves and may completely defoliate young stems. It can hardly be confused with any other pest.

There are species of sawfly other than N.*salicis* that may be expected at times to do damage. Inquiry is often made as to the cause of the red and green, globular or bean-shaped swellings that are so common on the leaves of willows. The insects responsible are certain species of gall sawfly, of which *Pontania gallicola* is the most common. The sawfly larva lives in the gall, and when full-fed burrows out and pupates in the soil or in cracks in the willow stumps. Unless the galls are numerous enough to weigh down the rods, the injury does not seem to be serious. Mites of the genera Phyllocoptes and Eriophyes deform willow catkins and cause hypertrophy of the reproductive organs. The floral bracts are enlarged, adventitious buds developing at their bases, the whole forming a 'witches broom' type of gall.

Insects under Suspicion

'Speck' or 'fleck' are terms applied by the trade to a specific form of timber damage. It is suspected of being due to the activities of one or other of the following two insects, the dipterous cambium-miner *Dizygomyza carbonaria,* the larvae of which tunnel in the cambium, the tunnels being covered over by successive wood rings, leaving specks or marks in the wood; another is *Dizygomyza barnesi,* which tunnels in the cambium of both basket willow and the cricket bat willow. The infected stems are often attacked by bacterial rots which induce cankerous growths. 'Speck' or 'fleck' damage is widely distributed and is more commonly found where the soil is sandy, although the defect is not inevitable on sand, as perfectly clean trees are often produced on soil of this type.

Thick covering of white scales on the bark of stems and branches. Female scales are broad and pear-shaped, those of the male being elongate and parallel-sided with a central raised keel. Its presence is indicative of ill-health of the tree due to other fundamental causes.	The Willow Scale (*Chionaspis salicis*)

Methods of Control

No suggestions as to practical methods of control have been given in the preceding sections, since the few forms of treatment known may be applied to meet attacks by several kinds of pest. Notes under this heading may be divided into (i) direct measures, which may be adopted when an attack is expected or is actually in progress; (ii) indirect measures, which will tend to reduce the numbers of the various pests and so prevent further attacks.

(i) Direct Measures—as a matter of general principle, when the foliage of any plant is eaten by beetles or caterpillars, the first measure to be considered is some form of poisonous spray that will leave a coating of poison on the leaves and so kill the insects as they feed. The poison most commonly used for this purpose is lead arsenate, but it should be noted that the spraying of basket willows with lead arsenate has proved rather ineffective in Britain, probably because the leaves have so smooth a surface that the poison cannot stick to them. If this explanation is correct, the difficulty should be easily overcome; indeed, successful results have recently been obtained experimentally by the use of a proprietary spreader. Until further trials have been carried out, however, it is suggested that to meet all attacks by leaf-eating beetles, caterpillars or grubs, a wash containing nicotine and soap should be used. Insecticides of this nature have been found efficient in practice, and their application is treated below.

When plants are attacked by aphides or other insects that feed by sucking up the juices of the plant and not eating the solid parts, it is necessary to use a contact insecticide, that is to say, an insecticide that kills insects touched by it. There are several such washes, but the only one that seems to have been properly tested on willow is nicotine, which is undoubtedly the best contact insecticide known. It has the additional advantage of being quite efficient when used against leaf-eating insects, such as willow beetles. Its chief disadvantage is its expense which is counter-balanced in some degree by the number of different pests that it can destroy. Experiments with alternative washes, eg pyrethrum, have been carried out by the National Fruit and Cider Institute, Long Ashton, Bristol, and growers are recommended to get into touch with this station, which gives special attention to problems concerned with willows grown for osier production.

The amount of spray fluid required per acre varies from 1·8 to 2·7hectolitres (40 to 60gal) according to the size of the rods, and a man with a knapsack machine can spray from 0·4–0·8 hectares (1–2 acres) a day, according to local conditions. Among the osier beds of Somerset a proprietary nicotine and soap wash is sometimes used, but those who wish to make up their own wash with a known nicotine-content might try the following formula, varying it from time to time to find the minimum percentage of nicotine that is effective:

Nicotine, 98 per cent	85g (3oz)
Soft soap	0·90–1·8kg (2–4lb)
Water	1·8hl (40gal)

Sufficient soap should be used to obtain a wash that 'lathers' well and it may be necessary with some waters to increase the quantity recommended in this formula. The number of times that it is necessary to spray is also variable; sometimes as many as three applications are made as a regular routine, but it is probable that if the willows are well sprayed in late May or early June it will seldom be necessary to spray again the same year.

Apart from spraying, which will deal with aphides, leaf-eating grubs and beetles, and to a smaller extent with the moth caterpillars that live in the shoots, no other direct measures of control can be recommended from actual experience. On the Continent the leaf-eating willow beetles are caught by various forms of apparatus, by means of which the insects are shaken off into trays, and some such method might be of service in this country in dealing with the Willow Weevil (*Cryptorrhynchus lapathi*) against which sprays are useless. No control measures are known for the midges causing 'button top' or for the various insects that burrow in the stumps or rods. Since, however, the insects causing 'button top' normally pass the winter in the buttons, and as affected rods are usually of little value, such rods should not be left about in the neighbourhood of the beds, or the midges will emerge and attack the new crop.

(ii) Indirect Measures—in addition to the direct measures of control that can be applied to the growing crop, there are certain precautionary measures that are worth consideration, as they tend to prevent the necessity for spraying. It has been pointed out that willow beetles spend the winter in heaps of rubbish under bark, etc. It is obvious, therefore, that all such heaps left until the spring are sources of danger, while equally, if they are burnt during the winter, they will have acted as traps and allowed the easy destruction of many pests. In the same way it is probably a mistake to allow old, pollarded willows—however picturesque—to remain close to willow beds, individual willow trees or ornamental trees. They are always thoroughly infested by willow-feeding insects of all kinds, and must act as centres from which these insects spread to healthy trees and plants. If willow trees are needed they should be of a variety that will pay for proper attention, and they should not be pollarded. Neglected and decayed willows, poplars, or alders, are all undesirable in the neighbourhood of cultivated willows. Partly dead, or weak willow stumps are usually attacked by insects that burrow inside, such as the larvae of the willow weevil and of the clear-wing moths. Such stumps should be removed and burnt. If they are left, the insects will emerge and make fresh attacks on both stumps and trees.

Problems of Unknown Origin

A number of pathological phenomena develop in willows, the cause of which is at present unknown. The most serious of these is the defect known as 'stain', 'bar stain', or 'butterfly mark'. When present in timber used for the manufacture of cricket bats it appears as a dark, horizontal bar across the bat; brown wings of rather lighter stain go out from the sides of the bar. In a standing tree the position of any large patch of stain in the wood is shown by the presence of a swelling or slight burr on the bark. This burr is usually cracked horizontally. If the stem is suitably split and cross cut the stained wood shows as a sector of the transverse section but it extends only for a short distance up and down the tree. Sometimes the injury is more severe, the cambium being killed instead of being stimulated to form abnormal wood. In this way hidden cankers arise, the presence of which cannot usually be recognised from the outside. Until the cause of stain is known with certainty, control measures cannot be suggested with confidence. It is most important to avoid pruning wounds on the stem of the set. Wherever rough bark forms unduly early and especially in small patches around buds, twigs or wounds, stain is to be expected. Rough bark is usually, if not always, caused by frost and this is to be countered chiefly by good drainage and stimulation of growth.

8 : *Diseases*

FUNGAL DISEASES

WILLOWS ARE NOT immune from attack by fungal diseases. By far the most important parasite in this group is the well-known 'honey agaric' *Armillaria mellea* (Vahl ex Fries) Kummer, whose hosts are numerous, ranging from herbs to forest trees. While infected trees do not invariably die, the death rate is high. After killing the tree the parasite thrives on the stump left in the ground, but as the organism cannot maintain itself if severed from its natural food base, root extraction becomes an essential feature of the process of eradication. Moreover, in the course of this procedure it is highly desirable that all, even the smallest piece, of infected root or stem should be collected and destroyed.

The first symptoms of attack are thinning of the foliage with premature autumnal colouring and leaf-fall. In large trees these effects may appear for two or more seasons before the tree succumbs. Conversely, in the case of shrubs, bushes, stools and recently-planted sets, the leaf symptoms appear for only a few weeks, whereupon the whole plant suddenly dies. Woody plants adjacent to the original infection may also become attacked and a slowly widening patch of dead plants is a typical indication of the presence of Armillaria in the locality.

Another proof of its presence may be the emergence of toadstools—the fruiting bodies of the fungus. They occur usually in the late summer or autumn around the bases of relatively large trees which have been killed by the infection. The toadstools produce myriads of spores which may be windborne to newly exposed stumps. The spore germinates to produce a mycelium which spreads under the bark, into the outer tissues of the wood and eventually into the dead roots. Here it forms rhizomorphs which, with the rotting of the bark, become exposed to the soil into which they grow for varying distances. Thus they become linked to their food supply in the rotting wood and can, under such conditions, remain alive for many years, forming a reservoir of infection and a threat to nearby trees. No commercially-grown tree or bush in Britain, nor, presumably, elsewhere, is known to be immune from Armillaria.

The organism attacks living trees only through the roots and by the time the tree shows symptoms of attack it is incurable. Obviously, the best way to reduce losses is to avoid planting where soil infection might be expected. For an existing attack of this fungus, the following method of eradication of the infection from the soil has been proved effective. Particularly is this so when used after removal of a diseased tree, or after removal of a tree that appears healthy at felling but is liable to suffer invasion by the fungus so long as the stump roots remain alive. The following instructions are for preparation of the site after clearing the infected tree and for subsequent fumigation with carbon disulphide. Incidentally, while the efficacy of this method is undisputed, its use in Britain is only on a meagre scale, mainly because of the time occupied which, of course, can be substantial if a large area is involved. In N America and Canada the method is more commonly used and with good effect.

THE METHOD

1. Fumigation is more effective in light than in heavy soils, and more effective in warm, dry soils than in cold, wet ones. It is therefore best carried out during a dry spell in summer months.
2. All roots, or as many as possible, including tap-root and

main laterals, should be removed from the soil and burnt.

3. When the roots have been cleared, the soil should be restored and levelled.
4. Apply carbon disulphide over the entire root area of the former tree, at a rate of 56ml (2 fluid oz) per charge every 45·72cm (or 18in), in rows also of 45·72cm apart. Charges should be staggered in adjacent rows, to make a triangular or diamond pattern. Apply charges 20cm (8in) deep in sandy soil, and 15cm (6in) deep in heavier soils. Injection will be facilitated if a self-measuring, force-feed injector is used.
5. Apply 4 depth-charges of carbon disulphide around the former site of the tap root; the rate is 170ml (6 fluid oz) per charge, 1·21m (4ft) apart and 1·2–1·8m (4–6ft) deep. A soil auger can be used for making the holes, and a suitable tube inserted to the right depth for delivery of the fumigant.
6. Tamp the soil well after each injection to prevent rapid escape of the fumes.
7. After the injections have been applied to the site, flood the soil to a depth of 7·62cm (3in) and maintain this blanket of wet soil for 3 weeks to prevent escape of the gas. Canvas sheets or tarpaulins can be used instead of the 'water seal' if more convenient.
8. At the end of 3 weeks, the soil can be allowed to dry out and the remaining gas to escape; replanting is said to be safe when the odour of carbon disulphide can no longer be detected.
9. All the above precautions must be observed if success is to be ensured; especially important, but frequently neglected, is the choice of suitable seasonal and weather conditions (see 1. above).
10. Observe the same precautions for storing and handling carbon disulphide as for petrol.

The maximum persistence of the organism in the soil is not known, but it can remain alive for as long as twenty years. If the above or any other proved method has not been carried out and it is decided to take the risk of replanting, a very close watch should be kept on the new plants.

Several other slightly less troublesome fungal diseases can attack willows and a few of these are listed below. One in particular—*Marssonina salicicola*—often incurs much damage among weeping willows and other ornamental types. It causes leaves to become discoloured, curled, and distorted, and finally to rot and drop off. *Fusarium lateritium* working in conjunction creates tiny cankers on the leaf stalks. Affected trees look decidedly unhealthy and in extreme cases become almost totally defoliated. The most effective treatment is to spray with Bordeaux mixture three or four times in the spring at fortnightly intervals. Alternatively, *Fentinhydroxide* might be used. A sickness of any kind has a tendency to weaken the constitution, and trees, like humans, need rather special feeding after an attack. In this instance a slow-acting compound fertiliser such as bone meal mixed with dried blood is considered effective, and should be lightly forked into the ground round the base of the tree. If the tree is in grass, it will be best to drill a few holes in the turf just around the periphery of the branches and fill these with fertiliser.

The fungal disease—*Physlospora miyabeana*—commonly known as the Black Canker of willows, is responsible for leaf-spot and stem canker. In England it is confined almost exclusively to the osier type or basket willows, especially *vitellina* and *triandra*. In N America and Japan, willows of various kinds are frequently attacked, resulting in much damage. This fungus is a true parasite, the germ tubes penetrating the outer skin or cuticle. After attacking the leaf, which blackens and droops, the fungus often grows down the petiole into the stem, discolouring it and forming a canker. It may also invade the ends of young twigs, causing them to shrivel and bend over. Osier rods when cankered are useless for basket making and should be destroyed. Cricket bat willow trees are rarely attacked, but stool-beds are certainly not immune. To prevent infection of osiers and bat willow stools they should be sprayed with Bordeaux mixture—once in very early spring and again a few weeks later.

Another somewhat similar disease which frequently works alongside *P.miyabeana* but developing a little earlier in the season, is *Fusicladium saliciperdum*. This is sometimes referred to as the 'scab' of willow and develops in pustules on the leaves, especially under the veins, and similar pustules on the twigs are a means of over-wintering. Yet another similar infection, sometimes known as 'bark scorch', occurs in willows on the Continent and in N America. It is *Venturia chlorospora* and in a wet season infected trees may be almost defoliated. In dry weather, however, it ceases to spread.

Another canker creating fungus is *Cryptomyces maximus*, which produces large black patches on willow branches, especially *S.fragilis*. Sometimes the branches of the cricket bat willow are similarly attacked, but severe attacks on tree willows in Britain appear to be of rare occurrence. In practice the diseases of *Fusicladium saliciperdum*, *Marssonina salicicola* and *Cryptomyces maximus* are so very difficult to distinguish one from the other that a great deal of confusion between them often occurs.

A number of rust fungi of the genus *Melamspora*, attack willow leaves, producing small yellow pustules which cause the leaves to wither and fall. It can be a serious menace if not checked, but the only known control is to spray with Bordeaux mixture during the growing season.

Trees which are in a weakened condition through any cause whatsoever are liable to attack by the fungus *Cytospora chrysosperma*, which will colonise and rapidly kill them. The fungus produces minute black spots (fructifications) buried in the bark, from which orange-coloured spore-threads are extruded, especially in damp weather during the later months of the year. As there is no evidence that it is ever a virulent parasite on vigorous trees, the removal of unfavourable conditions, eg overcoming the effects of drought by copious watering, is the only effective antidote. Another condition with which this fungus is sometimes associated is the bacterial infection *Erwinia salicis*, or Watermark disease (q.v.).

Capnodium salicium is the 'sooty mould' which grows in the honey-dew deposited by aphides on the leaves of willow. Its mycelium is superficial and harmless.

Only very rarely are tree willows attacked by *Stereum purpureum*, the Silver Leaf disease, which is more commonly associated with fruit trees and several forest hardwoods. Unlike these latter trees, willow rarely, if ever, exhibit the 'silvery leaves' characteristic of this disease, which, when observed in the other trees, is due to the fungus in the woody parts below. The fungus is not present in the leaves. It also occurs commonly as a saprophyte on recently exposed stumps of willow and other trees. Attacked trees do not invariably die, but speedily lose their value.

Willows grown on very short rotation, which is common to the cricket bat variety, do not as a rule suffer from heart-rot. Trees of about eighteen years and over, however, have a tendency to develop this, and in such cases it is found to extend into the trunk from either crown or root. The fungi causing heart-rot in pollards and old unpollarded trees may be either *Trametes suaveolens*—a fungus whose fructification, when fresh, smells of aniseed, though seldom recorded in Britain—*Fomes igniarius, Polyporus sulphureus*, or a wound parasite called *Ganoderma applanatum*.

Occasionally willows are found with heart-rot halfway up the trunk, with both top and bottom remaining sound. Investigation usually reveals that the trees are excessively tall through being drawn up by their too-closely-planted neighbours, and in the absence of proof to the contrary it is surmised that the heart-rot in the half-way situation may be due to this same reason.

Apparent ill-health symptoms of willows may be due to *Cryptodiaporthe salicina,* which is thought by some to be a doubtful pathogen but has been found sporing freely on cricket bat willow stools. *Rhytisma salicinum,* or Tar Spot, is responsible for jet-black leaf spots, very definitely bounded, about 0·6cm (¼in) diameter and somewhat raised above the leaf. A severe attack can weaken the tree through defoliation.

Other leaf-spot damage of a less serious character can also be caused by *Rhytisma symmetricum* (seldom noted in Britain), *Scleroderris fuliginosa* and *Septoria salicola.* The general antidote to the majority of these fungal problems is to spray once or twice during the spring with Bordeaux mixture.

BACTERIAL DISEASES

The Willow Twig Blight, *Pseudomonas saliciperda,* is usually found on the ornamental species of the genus and principally, but not entirely, in New England, USA, and occasionally in Great Britain. It has the effect of turning the leaves brown and making them wilt while any branches affected usually die back for some distance. The general effect of an attack on the leaves of young trees and shrubs is to make them appear as frozen. In fact, the attack is often confused with frost damage. Also, brown streaks sometimes appear in sections of the wood. As the parasite winters in cankers the young leaves tend to become infected as soon as they unfold. One recommended control measure is to cut away and burn all infected twigs, but as an alternative the plant may be sprayed two or three times during the early spring with Bordeaux mixture.

By far the most devastating infection among willows is the Watermark disease, the causal bacterium of which is known as *Erwinia salicis.* Its importance stems from the fact that for the past fifty years it has wrought much havoc among the chief timber producing willows—the coerulean or cricket bat variety. Fortunately, through the good offices of the local authorities of Essex, Suffolk, Norfolk, Cambridgeshire, Bedfordshire and Hertfordshire, and supported by Statutory Orders implemented by the Forestry Commission, the scourge has been brought under a large measure of control. Excellent as these efforts have been, and are continuing to be, complete eradication has not been achieved; constant watchfulness, and action where necessary, is still required.

While the origin of Watermark disease is a little obscure, the pathogen is believed to have reached Britain a little over half a century ago from the European continent—probably Holland or Austria, where a disease confined to the willow genus and having almost identical characteristics has been known for a great number of years. It was first discovered in Britain in the White Willows of Cambridgeshire and in the mass of open-bark *fragilis*-type to be found along the Norfolk bank of the River Ouse from which parts it spread to other areas of southern England, particularly the south east. Its chief significance lies in the fact that if not kept within some measure of control it could quickly place the game of cricket in jeopardy through restricting the supply of first grade willow timber for the manufacture of bats.

Although the disease had been observed from a very much earlier date, it was not until the early 1920s that the causal germ was isolated and found to be a bacterium, rod-like in shape and possessing flagella used as a means of propulsion. It was eventually named *Erwinia salicis*.[4]

The causal germ lives in the water-containing vessels of the wood and, during the rise of sap in early spring, becomes distributed over large areas of the tree. With the rapid increase in bacteria a complete blockage of the sap vessels takes place in many of the shoots bringing about the death of the newly-expanded foliage which then withers and turns a reddish-brown colour. This red-leaf is the earliest and most significant outward symptom of the disease and usually signifies that much has been happening within the tree over many months and not that the infection has necessarily occurred only a short time before. The leaves on any one twig die as a whole, but the dying back on each branch is usually progressive so that the leaves wither on one part after another. The crown is rarely affected as a whole—part only dying back as one or more branches become affected. They may die back at one time or in succession and observations show that no die-back takes place after July, either on trees already diseased or on those hitherto immune. In early autumn some branches on diseased trees exhibit a premature leaf-fall, beginning with the lower leaves on each twig, and this is the last external symptom of disease observed during any one year.

The bacteria continue to spread within the tree, but the rate of spread becomes slower as the sap in the branches becomes less. In addition to 'red-leaf' and 'die-back', two other symptoms can occasionally be observed. Adventitious shoots, or secondary growth, appear along the lower branches of the affected tree and, also, an opaque liquid exudes from small wounds and insect exit-holes, on the dying twigs. These exudations have their source in the mass of bacteria inhabiting the vessels and, unless used as a habitat by fungi or washed away by rain, the mass gradually sets hard and takes on a yellow or brown colour on the trunk and branches.

Within the tree itself the chief distinguishing mark is the watermark stain on the timber and it is from this that the infection derives its name. The stain is found when cutting through a twig or branch having either dead or still-living wood recently infected.

In its first stages the watermark appears throughout the dying branches, later progressing into the main stem of the tree. It extends upward to some degree, but much more so, and more rapidly, in a downward direction, and generally, but

Page 155 (left) The hybrid, S. fragilior, *often reaches large dimensions, is quite beautiful, and resembles a forest tree more than any other form of willow. Appropriately, it is known as 'the Monarch of the Willow family'*

(Right) *Another progeny of* alba × fragilis, *and a variety of* S. viridis. *It is* S. elyensis, *first discovered at Ely in Cambridgeshire*

Page 156 (left) *Defoliation by caterpillars;* (right above) *an ill-fated willow. Evidence of heart-rot, Honey fungus, Watermark disease and a severe attack by frost;* (right below) *a further symptom of Honey fungus: whitish mycelium spreading under the bark*

not always, enters the root system. It occupies the one year old annual ring and, to a slight extent, the autumn wood of the older adjoining ring. In the dying branches the current year's ring is also occupied. Cases have been known where the watermark has occupied the entire wood, even at this stage.

While in the smaller branches the whole of the annual ring is usually watermarked, one side only may be affected in the larger wood. As the watermark spreads upwards or downwards it follows a twisting course in an anti-clockwise direction (when looking at the roots). Within the root-system itself the development of the watermark is similar to that within the crown. Once a branch becomes infected the disease persists from year to year, successive outer rings becoming invaded until the branch succumbs. It is quite clear that the infection of an outer annual ring takes place from an inner ring and not from an external source. If no infection of the new wood occurs all succeeding annual rings are free from infection; as the branch thickens the diseased part is buried and the branch may then recover. It can, of course, be infected afresh from an external source, whereupon the earlier staining system is repeated.

The wood of recently infected trees is usually sodden with sap-coloured bacterial slime which, upon exposure to the air, becomes darker in colour and is characteristic of the disease. When a tree which has been diseased for two or more years is cut down, it will be seen that the outer portion of the circular watermark stain is only faintly coloured, while the stain in the inner or earlier rings varies between dark brown and black, denoting that the disease has ceased to spread in those parts. It is in the outer and practically colourless portion that the parasite is living and active; in the inner and darker parts it is usually dormant. Where the wood has, in fact, been dead for some time the stain fades to a light greyish colour and it is believed that the bacteria then quickly die, although there is no conclusive proof of this.

The final stage is, of course, the death of the tree, which rarely occurs as a direct effect of the watermark bacteria. From an economic point of view this stage is relatively unimportant as once the tree is attacked its value is considerably reduced, and under present arrangements where Statutory Orders apply it is compulsorily destroyed. If the tree does succumb through the effect of the disease the watermark is always present in the current year's wood on the main stem.

While *Erwinia salicis* is the germ responsible for Watermark disease, its invasion of a tree is quickly followed by associated organisms requiring an environment such as is created by the pathogen. These organisms, known as saprophytes, are incapable of feeding on live tissue and find suitable nourishment in the dead and dying leaves resulting from the blocking of the sap vessels—referred to earlier. They soon find their way right into the tree, following the course of the watermark and living on the sap changed in character by the activities of the pathogen. They then proceed to stain and dissolve away certain important woody elements, resulting in minute cracks and fissures being formed in the timber. The effect of this disintegration of woody elements usually becomes apparent when the timber is undergoing compression during the process of manufacture. If, perchance, the cleft withstands the compression it is more than likely that the weakening effect of the disease will be manifested unexpectedly on the cricket field, with the impact of the ball on the blade. The importance of the disease, and the secondary bacterial invasion in particular, should not, therefore, be underrated. A further saprophyte, a fungus known as *Cytospora chrysosperma*, while making dead tissue its prime host is also able to attack and kill live tissues which have been weakened by the presence of other diseases. If it attacks a watermarked tree in early summer—as frequently happens—the tree succumbs much sooner than would otherwise be the case and by autumn or winter the bark is thickly studded with the yellow thread-like fructifications peculiar to the fungus.

Watermark disease is highly infectious and insects and birds are believed to be the chief agents for its transmission and distribution. The feeble-flying insects carried on the wind are the first suspects, as the spread of disease is found to be more extensive in the direction of the prevailing wind than otherwise. Among other suspected carriers are the Willow Gall Midge, Willow Sawfly, Willow Wood Wasp, Goat Moth, several species of weevil and small birds, especially the Willow Tit. The latter, when seeking grubs, often rips open shoots and, should the disease be present, the bird may carry the causal bacterium on its beak and introduce it into shoots on healthy trees.

As the opaque liquid, mentioned earlier, contains large quantities of the bacteria this also can spread the disease, if, perchance, it falls into wounds on nearby trees. This is a strong point in favour of not planting willows so closely together that their branches overlap when fully grown. In fact, the planting of trees too closely together is a factor contributing to the spread of disease, chiefly through the meeting and intermingling of branches, resulting in some trees becoming dominant and tending to suppress their neighbours. When this occurs it is the suppressed trees, and those with heavily shaded branches, that are the first to show disease symptoms. A minimum planting distance is 11m (35ft) for trees in single line and 13m (40ft) in plantations.

Over the years the general opinion has been that infection takes place only in the crown of the tree, but I have established beyond doubt that trees can also be infected through their roots as a result of the bacteria being carried in water seeping through soil or flowing down-stream.

Of the many types of timber willow indigenous to this country none are known to be resistant to Watermark disease, and this applies equally to the several strains of cricket bat willow. If, indeed, a resistant strain could be found, and at the same time its timber proved to be suitable for bat manufacture, the whole problem would at once be resolved. No practical means are known for protecting trees from attack, although preventive measures can to some extent be tied up with cultivation factors. For instance, sites having marshy tendencies with a definitely wet sub-soil or a water table constantly just beneath the surface are known to favour the spread of disease.

Investigations carried out in recent years have revealed that the old pollarded open-barked type of tree, usually of the species *S.fragilis,* the crack willow, and prolific in many parts of the country, is a ready host to the watermark pathogen. Many such trees when felled not only contain the internal stain symptom of the disease but also bear indications that the infection has been present for a number of years. The fact that the 'red-leaf' outward symptom is seldom produced in such trees adds considerably to the problem of detection. Also, owing to the peculiar systematy of the crack willow, difficulty is experienced in finding the disease stain when cuttings are taken—as many as a dozen or more cuts having to be taken before the stain is found. The destruction of these trees, which are usually large, often presents a financial problem, as frequently they are situated in places difficult of access, consequently increasing the cost of clearance. Nevertheless, their destruction is vital if the spread of infection is to be held in check.

The use of insecticide sprays as a preventive measure is a possibility, but scarcely a practical proposition owing to the heavy cost involved and the widespread areas of the cultivated trees. Moreover, the inevitable killing of birds resulting from such wholesale methods is greatly to be deplored. Despite the fact that a few birds may unwittingly become vectors of the causal bacteria, the damage inflicted can scarcely be compared with the help rendered by the birds in clearing immense numbers of caterpillars and other insects, thereby preventing much defoliation and twig damage to trees and plants in general. To spray them with lethal chemicals seems a poor reward for such excellent labours.

Glossary

ABORTIVE: imperfectly or not correctly developed, barren.
ACUMINATE: tapering at the end; long pointed.
ACUTE: sharp pointed; ending in a point.
ALPINUS: belonging to mountains.
ALTERNATE: not opposite on the axis.
AMENT: the catkin.
ANDROGYNOUS: both staminate and pistillate flowers in same inflorescence.
ANGIOSPERMOUS: the seeds borne within a pericarp.
ANTHER: pollen-bearing part of stamen.
APICULATE: contracted into a minute point.
APPRESSED: lying close and flat against.
ARTICULATE: having a node or joint.
ASCENDING: rising obliquely and curving upward.
ATROVIRONS: dark green.
ATTENNATE: scantily tapering.
AURITA: eared.
AWL-SHAPED: tapering from base to small but stiff point.
AXIL: upper angle formed by leaf or branch with stem.

BI- OR BIS-: Latin prefix for two or twice.
BIFID: two cleft.
BINOMIAL: providing for two names.
BILABIATE: two-lipped.
BISEXUAL: having both stamens and pistils.
BLADE: expanded portion of leaf.
'BLOOM': (see 'glaucous').
BOREALIS: northern.
BRACT: leaf-like structure at base of leaf.
BRACTEOLE: small bract on secondary axis.
BRANCHLET: small branch of either current or preceding year.

CAESIUS: blue-green.
CALYX: outer whorl of flower, comprised of sepals.
CANESCENT: grey-pubescent and hoary.
CAPSULE: hard, dry pod holding seeds, which when ripe are shed through valves, slits or pores.

CARPEL: seed-producing part of flower; whole or part of ovary.
CARPELLARY: relative to female flower.
CATKIN: mode of flowering of willows; inflorescence of cluster of small wind-pollinated flowers, usually unisexual.
CELL: minute vesicle of which plant tissues are formed; cavity of anther or ovary.
CILIATE: fringed with hairs—as with eyelashes (leaves).
CINEREA: ashen grey.
COERULEA: heavenly blue.
CONCOLOROUS: of uniform tint.
CONNATE: united; bases of opposite leaves joined together.
CORDATA: heart-shaped.
CORIACEOUS: of leather or leather-like texture.
COROLLA: inner part of flower, comprised of petals.
COTYLEDON: primary leaf or leaves in embryo.
CRENATE: toothed with rounded, shallow teeth; bluntly scalloped.
CRENULATE: minutely crenate.
CUNEATE: wedge-shaped, narrow end at point of attachment.
CUSPIDATE: tipped with sharp, rigid point or cusp.
CYLINDRIC: hollow roller-shaped (of stalks).

DAPHNOIDES: laurel-like.
DECIDUOUS: not persistent; sheds leaves in due season.
DECORTICATED: deprived of bark.
DECUMBENT: reclining but with ends ascending.
DEFLEXED: bent outward, opposite to inflexed.
DEHISCENT: opening to emit contents, as capsule or anther.
DENTATE: toothed with teeth directed outward.
DENTICULATE: minutely toothed.
DEPRESSED: flattened from above.
DIOECIOUS: having flowers of separate sexes on separate plants.
DIVARICATE: widely spreading apart.

ELLIPTIC: shaped like an ellipse.
ELLIPTIC, ELLIPSOID: ellipse, regular oval.
EMARGINATE: with a shallow notch at the apex.
EMBRYO: the rudimentary plant within the seed.
ENTIRE: leaf margin even, not toothed or lobed.
EPIPHYTIC: growing on other plants, but not parasitic.

FALCATE: sickle- or scythe-shaped.
FASTIGATE: with stems or branches erect and near together.
FERRUGINEOUS: rust-coloured.
FERTILE: capable of producing fruit, seeds or pollen.
FIBRO-VASCULAR: composed of woody fibres and ducts.
FILAMENT: stalk of an anther.
FILOSE: terminating in a thread-like manner.
FLORIFEROUS: flower-bearing.
FRAGILIS: brittle.
FRUIT: seed-holding structure; in willow a two-valved capsule.
FRUTICOSE: shrubby; with woody persistent stems and branches.
FUNICLE: stalk of an ovule or seed.
FURROWED: with longitudinal channels or grooves.
FUSCOUS: greyish-brown.

GENERIC: characteristic of a genus or class.
GENUS: group of plants comprising a number of species of common character.
GERMINATE: sprout, bud, cause to shoot or produce.
GLABRATE: nearly glabrous.
GLABRESCENT: nearly glabrous with age.
GLABROUS: not hairy, smooth.
GLAND: fleshy structure on stem or leaf.
GLANDULAR: bearing glands or gland-like cells.
GLAUCOUS: whitish- or bluish-green caused by 'bloom'.
GLOBOSE: globe-like, nearly spherical.

HABITAT: natural home of plant.
HELIX: crooked.
HOARY: covered with close whitish or greyish-white pubescence.
HOMOGAMOUS: bearing only one kind of flower.
HUMUS: the brown or black constituent of the uppermost layer of soil, consisting mainly of organic compounds, resulting from the decay of plants and animals left behind in the soil.
HYBRID: plant resulting from a cross between two or more plants more or less unlike.
HYPOGYNOUS: situated below pistils or ovary.

IDENTIFY: establish identity or individuality of.
INCANA: pale grey.
INDUMENTUM: hairy covering (leaves).

INFLORESCENCE: mode of flower-bearing.
INTEGERRIMA: entire-edged.

LANCEOLATE: lance-shaped, about four times as long as wide and broadest below or about middle.
LIGNEOUS: woody.
LINEAR: long and very narrow, sides almost parallel.
LUNATE: half-moon or crescent-shaped.

MEDULLARY: pertaining to pith or medulla.
MONAECIOUS: with unisexual flowers of both sexes on same plant.

NECTARY: place or organ where sugar or nectar is secreted.
NIGRA: black.
NODE: place upon stem which normally bears leaf or leaves.
NOMENCLATURE: naming of plants; scientific naming is binomial —generic and a specific, eg *Salix* (generic), *alba* (specific).

OB-: Latin prefix usually signifying inversion.
OBLANCEOLATE: lanceolate, but broadest near the apex and tapering from that point.
OBLIQUE: unequal-sided.
OBLONG: about three times as long as wide and with nearly parallel sides.
OBLONG-OBLANCEOLATE: oblong from the apex, then tapering to stem.
OBOVATE: inversely ovate; rounded apex, tapering to stem.
OBOVOID: inversely egg-shaped.
OBTUSE: ovule-bearing part of pistil.
OVATE: having outline like hen's egg.
OVULE: body which after fertilisation becomes the seed.

PALUSTRIS: belonging to swamps.
PATENT: protecting (hairs on ovary).
PEDICEL: stalk of flower.
PEDICELLATE: borne on a pedicel.
PEDUNCLE: fruit or flower stalk.
PEDUNCULATE: borne on peduncle.
PENDULOUS: hanging; weeping.
PENTANDRA: five-stamened.
PERICARP: wall of ripened ovary.

PERSISTENT: remaining attached; not falling off.
PETIOLE: leaf-stalk.
pH VALUE: an important feature in choosing the most suitable soil; denotes content of hydrogen ions per unit volume of soil; chemically pure water is neutral or is given a pH of 7; a lower reading indicates a more acid soil, a higher reading more alkaline: a pH value of 6·5 is most suitable for willows.
PILROSE: soft, long, straight hairs.
PINNATE: (leaf) compound, with leaflets on each side of rachis.
PISTIL: complete seed-bearing organ of flower.
PISTILLATE: having pistil and no stamens.
PLUMROSE: feathery, furnished with long hairs.
POLLARD: tree cut back usually to about 2m (7ft).
POLLEN: spores or grains borne by anther, usually granular.
PROCUMBENT: lying on ground or trailing.
PROSTRATE: lying flat on ground.
PRUINOSE: bloomy.
PUBERULENT or PUBERULOUS: minutely pubescent.
PUBESCENT: covered with hairs, particularly if short and soft.
PUNCTATE: bearing translucent or coloured dots or depressions.
PYRAMIDAL: shape of figures (trees) having sloping side rising from flat base and meeting at or about apex.

RACHIS: leaflet or flower-bearing axis.
RECUMBENT: lying or reclining.
RECURVED: curved downward and backward.
REFLEXED: abruptly turned downward.
REPENS: creeping.
RETICULATE: divided in fact, or appearance with network of veins.
REVOLUTE: (leaf) margin rolled backward toward lower side.
RIB: (leaf) primary or prominent vein.
RODS: willow stems grown as osiers.
RUBRA: red.
RUFOUS: reddish brown.
RUGOSA: wrinkled.
RUST: fungal disease on leaf.

SALICIFOLIA: willow-leaved.
SCALE: protective leaf-like structure or bract.
SEPAL: outer leaf-like member of flower.

SESSILE: not stalked.
SHEATH: a tubular envelope.
SHRUB: wood plant branched from base.
SILKY: covered with appressed, fine and straight hairs.
SMOOTH: without roughness; glabrous.
SPECIES (sing and plur): stage in plant classification next below genera.
SPECIFIC (name): term now used for the 'trivial' of Linnaeus.
STAMEN: male organ of flower, bearing pollen.
STEM: main axis of a plant.
STIGMA: receptive tip of style which catches pollen.
STIPITATE: having a stalk.
STIPULE: small leaf-like organ(s) at base of petiole.
STRIATE: marked with fine longitudinal lines.
STYLE: projecting organ of carpel.
SUB-SHRUB: an undershrub; small low shrub.
SUBULATE: awl-shaped.
SYNCARPOUS GYNOECIUM: united female parts of a flower.

TAPERING: becoming gradually smaller in diameter of width toward one end.
TOMENTOSE: with dense woolly pubescence.
TOMENTULOSE: closely and finely tomentose.
TOMENTUM: dense covering of matted hair.
'TRIVIAL': term used by Linnaeus for the second name in his binomial system of plant naming—now designated as the 'specific'.
TORTUOUS: twisted or bent.
TREE: a woody plant with one main stem and at least 4m or 5m (12–16ft) tall.
TRUNCATE: ending abruptly, as if cut off.

UNDULATE: with wavy surface or margin.
UNISEXUAL: of one sex, either staminate or pistillate—(stamens or carpels).

VARIETY: plant differing from its species, difference produced either artificially or by natural causes.
VEINS: threads of fibro-vascular tissue in a leaf or other organ; nerves but especially the smaller, branched nerves.
VELUTINOUS: velvety.

VENATION: arrangement of veins.
VERNATION: arrangement of leaves, in bud.
VILLOUS: bearing hairs, long and soft, usually curved or curly.
VIMINALIS: osier-like.
VIRDIA: green.
VISCID: semi-fluid clinging consistence, like egg-yolk.
VULGARIS: common.

WOOLLY: having long and soft, and curled or matted, hair.

References

('Superior' numbers appearing throughout the text relate to numbers shown below).

1 Bean, W. J. *Trees and shrubs hardy in the British Isles* Vol III R–Z (1951)
2 Blackburn, Benjamin. *Trees and shrubs in eastern* N *America* (1952)
3 Curtis and Bansor. *Complete guide to* N *American trees* (1943)
4 Day, W. R. Oxford *Forestry Memoir* No 3 Watermark Disease (1924)
5 de Candolle, A. *Prodromus* (1868)
6 Edlin, H. L. *Wayside and woodland trees* (1964)
7 Edlin, H. L. *Trees, woods and man* (1956)
8 *Forestry Commission Bulletin* No 17 Third Edn (1968)
9 Harlow, W. M. *Trees of the Eastern and Central United States of America and Canada* (1957)
10 Linnaeus, Carl. *Species Plantarum* (1753)
11 Meikle, R. D. *British trees and shrubs* (1958)
12 Moldenk, H. N. & A. L. *Plants of the Bible*—Salix 216–8 USA (1952)
13 Raven, J. and Walters, M. *Mountain flowers* (1956)
14 Rehder, A. *Manual of cultivated trees and shrubs* New York (1958)
15 Seward, Sir A. C. *Plant life through the ages* New York (1959)
16 Smith, Sir J. E. *Flora Britannia* Vol 3; 1072 (1805)
17 Turrell, W. B. *British plant life* (1948)
18 Tutin, T. G. Edit. (K. H. Rachinger—Salix). *Flora Europaea* Vol 1 (1964)
19 Watson, H. C. *Cybele Britannica and Compendium* Vol 2 (1849)

Acknowledgements

The basis of preparation of this book has been my very close personal experience with the willow genus over a period of some 25 years. Even so, the work could scarcely have been achieved without reference to both early and contemporary writers. Any material used, however, has been limited strictly to that having proven reliability.

I wish to record my thanks particularly to the Cambridge University Press for permitting extracts from *Flora Europaea* Vol 1 (1964); to the Forestry Commission in London for the use of material from the Commission's Bulletin No 17, 3rd edition (1968), relative to the cricket bat willow; and to Herbert L. Edlin for permission to use extracts from several of his authoritative arboreal works.

Most useful guidance has also been received from R. D. Meikle, of Kew Gardens, whose views and opinions regarding botanical aspects of the willow genus are acceptable in all quarters. For this help I am grateful.

I would like also to thank a number of persons whom I know would wish to remain anonymous, but who have helped greatly in a number of ways.

Index

AT